Lecture Notes in Computer Science　　12474

More information about this series at http://www.springer.com/series/7412

Martin Reuter · Christian Wachinger ·
Hervé Lombaert · Beatriz Paniagua ·
Orcun Goksel · Islem Rekik (Eds.)

Shape in Medical Imaging

International Workshop, ShapeMI 2020
Held in Conjunction with MICCAI 2020
Lima, Peru, October 4, 2020
Proceedings

Springer

Editors
Martin Reuter (iD)
Harvard Medical School
Boston, USA

German Center for Neurodegenerative
Diseases
Bonn, Germany

Hervé Lombaert (iD)
Inria Sophia-Antipolis
Valbonne, France

ETS Montreal
Montréal, QC, Canada

Orcun Goksel (iD)
ETH Zurich
Zürich, Switzerland

Christian Wachinger (iD)
Klinikum der Universität München
LMU München
München, Germany

Beatriz Paniagua
University of North Carolina at Chapel Hill
Kitware Inc.
Carrboro, NC, USA

Islem Rekik (iD)
Istanbul Technical University
Istanbul, Turkey

ISSN 0302-9743   ISSN 1611-3349 (electronic)
Lecture Notes in Computer Science
ISBN 978-3-030-61055-5   ISBN 978-3-030-61056-2 (eBook)
https://doi.org/10.1007/978-3-030-61056-2

LNCS Sublibrary: SL6 – Image Processing, Computer Vision, Pattern Recognition, and Graphics

This Springer imprint is published by the registered company Springer Nature Switzerland AG
The registered company address is: Gewerbestrasse 11, 6330 Cham, Switzerland

Preface

This volume contains the proceedings of the International Workshop on Shape in Medical Imaging (ShapeMI 2020), held virtually in conjunction with the 23rd International Conference on Medical Image Computing and Computer Assisted Intervention (MICCAI 2020), on October 4, 2020, in Lima, Peru. ShapeMI 2020 is a continuation of the previous MICCAI ShapeMI 2018, SeSAMI 2016, and SAMI 2015 workshops, as well as the Shape Symposium 2015 and 2014.

Shape and geometry processing methods have been receiving significant attention as they are applicable in various fields from medical image computing to paleontology, anthropology, and beyond. While imaging is the primary mechanism to acquire visual information, the underlying structures are usually 3D geometric shapes, which often represent continuous or time-varying phenomena. 3D shape models, therefore, better describe anatomical structures than voxels in a regular grid and can have a higher sensitivity to local variations or early disease/drug effects relative to traditional image-based markers such as the volume of a structure. Therefore, shape and spectral analysis, geometric learning and modeling algorithms, as well as application-driven research are the focus of the ShapeMI workshop.

In ShapeMI, we strive to collect and present original methods and applications related to shape analysis and processing in medical imaging. The workshop provides a venue for researchers working in shape modeling, analysis, statistics, classification, geometric learning, and their medical applications to present recent research results, to foster interaction, and to exchange ideas. As a single-track workshop, ShapeMI also features excellent keynote speakers, technical paper presentations, and state-of-the-art software methods for shape processing.

We thank all the contributors for making this workshop such a huge success. We thank all authors who shared their latest findings, as well as the Program Committee members who contributed quality reviews in a very short time. We especially thank our keynote speakers, who kindly accepted our invitation and enriched the workshop with their excellent presentations: Ross Whitaker (Professor at The University of Utah, USA), Aasa Feragen (Professor at Technical University of Denmark, Denmark), and Stefan Sommer (Professor at University of Copenhagen, Denmark). We sincerely hope to meet you all in person at our next ShapeMI workshop.

October 2020

Martin Reuter
Christian Wachinger
Hervé Lombaert
Beatriz Paniagua
Orcun Goksel
Islem Rekik

Organization

Advisory Board/Program Committee

Aasa Feragen	DIKU, University of Copenhagen, Denmark
Claudia Lindner	The University of Manchester, UK
Diana Mateus	École Centrale de Nantes, France
Ellen Gasparovic	Union College, New York, USA
Ipek Oguz	Vanderbilt University, USA
James Fishbaugh	New York University, USA
Julia Schnabel	King's College London, UK
Kathryn Leonard	Occidental College, USA
Kilian Pohl	SRI International, USA
Marc Niethammer	University of North Carolina at Chapel Hill, USA
Martin Styner	University of North Carolina at Chapel Hill, USA
Miaomiao Zhang	Lehigh University, USA
Philippe Buechler	University of Bern, Switzerland
Shireen Elhabian	SCI, University of Utah, USA
Stanley Durrleman	Inria and ICM, France
Stefan Sommer	University of Copenhagen, Denmark
Steve Pizer	University of North Carolina at Chapel Hill, USA
Thomas Vetter	University of Basel, Switzerland
Tim Cootes	The University of Manchester, UK
Tinashe Mutsvangwa	University of Cape Town, South Africa
Umberto Castellani	University of Verona, Italy
Xavier Pennec	Inria Sophia Antipolis - Médi, France
Yonggang Shi	University of Southern California, USA
Yoshinobu Sato	Nara Institute of Science and Technology, Japan
Guido Gerig	New York University, USA
Ross Whitaker	The University of Utah, USA
Ender Konukoglu	ETH Zurich, Switzerland
Marius Linguraru	Sheikh Zayed Institute, UAE
Washington Mio	Florida State University, USA
Sungmin Hong	(MGH/HMS), USA

Contents

Applications

Methods

Composition of Transformations in the Registration of Sets of Points or Oriented Points

Jacob J. Peoples$^{(\boxtimes)}$ and Randy E. Ellis

Queen's University School of Computing, Kingston, ON, Canada
jacob.peoples@queensu.ca

Abstract. Registration of point sets in medical imaging applications may in some cases benefit from application-specific rather than general models of deformation by which to transform the model point set to the target. Further, including orientation data with the points may improve accuracy. To facilitate this, we propose an algorithm to register sets of points or oriented points through arbitrarily composed sets of transformations, so as to allow the construction of context-specific deformation spaces. The algorithm is generic with respect to the choice of transformations, requiring only that each constituent has a known solution to a particular standard form of equation. Our approach is framed in the mixture model framework, and constitutes a generalized expectation maximization algorithm. We present experimental results for two models—a 2D model of a cardiac ventricle, and a 3D model of a bug—testing the algorithm's robustness to noise and outliers, and comparing the accuracy when using points or oriented points. The results suggest the algorithm is quite robust to both noise and outliers, with inclusion of orientation data consistently resulting in more accurate registrations.

Keywords: Nonrigid registration · Expectation-maximization · Oriented points

1 Introduction

Point set registration is an important component of a variety of medical imaging tasks including computer assisted surgical navigation, and segmentation. Nonrigid registration is especially important for deformable anatomies such as soft tissues or bone complexes, or for registration with atlases, templates, or between individuals. For nonrigid registration there is, in general, a balance to be struck in order to ensure that the transformations are on the one hand, sufficiently flexible so as to capture real-world variations, while on the other, avoiding anatomically infeasible configurations. Application-specific models of nonrigid transformations may be a useful way to improve this balance: introducing functional or physical knowledge, *a priori*, into the solution, may allow more accurate specification of the space of deformations, thereby increasing specificity without sacrificing

© Springer Nature Switzerland AG 2020
M. Reuter et al. (Eds.): ShapeMI 2020, LNCS 12474, pp. 3–17, 2020.
https://doi.org/10.1007/978-3-030-61056-2_1

generalizability [26]. To facilitate the construction of such models, we consider a special case—namely complex transformations constructed by modularly composing simpler transformations—and propose a generic algorithm by which such transformations may be optimized to register two feature sets.

Composition of multiple transformations has often been a part of registration and segmentation methods, as well as anatomical or morphological models. Active shape models [6], for example, were fit to images by alternating between the optimization of pose—a rigid transformation—and shape, via the weights of a statistical shape model. Statistical human body models have often been composed of decoupled shape and pose models [2,15], as have models for segmentation of bone complexes such as the spine [23] or wrist [1]. For soft tissues, combinations of rigid transformations and biomechanical deformation models have been used for multimodal registration of prostate images [13]. This approach was made more accurate by including a statistical shape model of the prostate in addition to the biomechanical model [12]. In general, many objects, in medical applications and otherwise, are articulated—their pose being described by a hierarchy of basic transformations [11]—or can be better characterized through hierarchical models operating at multiple levels of detail [20,28].

Methods using composed transformations tend to require specific algorithmic approaches designed for the complete sequence of transformations being applied, complicating their use. An alternative is to simply use separate registrations, allowing each to refine the result of the previous with a transformation having more degrees of freedom [9]. Though this approach is common—at the very least, to initialize a nonrigid registration with a rigid or affine pre-registration [21]— the optimal result for earlier registrations in general will depend on lower-level shape variations induced by the subsequent transformations. Simultaneous optimization of all transformations may therefore be preferable.

We previously reported the use of a hierarchical, composed model of deformation for the segmentation of cardiac ultrasound [22]. We used a heuristic update and fallback strategy for the transformation during the iterative registration algorithm, which allowed the transformation update to be broken into a sequence of simpler optimization problems. Here we generalize the heuristic and fallback strategies to update arbitrarily composed transformations, requiring only that each component transformation has a known solution to a particular form of equation. This allows composition of transformations in a modular fashion, requiring no derivations specific to the particular chain of composition being employed. The algorithm, designed for use with sets of either points, or oriented points, is demonstrated on two models—a left cardiac ventricle, and a model of a bug—and the performance of points vs. oriented points is compared.

2   Mathematical Framing

Mixture models are a common framing in state-of-the-art point set registration methods [16]. Typically, one point set is taken to parameterize a distribution— namely a mixture model—of which the other is considered a sample. Solving for the transformation aligning the two point sets can then be treated as a maximum likelihood (ML) estimation problem typically solved using the expectation-maximization (EM) algorithm [7]. Broadly, EM algorithms iterate between a so-called E-step, wherein a posterior distribution for a set of hidden variables is estimated, and an M-step, wherein an optimization problem is solved to update the parameters. In this context, the statistical framing and the choice of transformation are largely orthogonal in that a given statistical parameterization can be used to register point sets through any transformation insofar as the M-step can be solved. Given this, and in light of recent interest in the use of oriented points in this area [3,5,18,22,24] we consider a generic formulation both in terms of choice of statistical distributions in the mixture, and the space of points.

Let capital letters such as $\mathbf{X} = \{\mathbf{x}_i\}_{i=1}^{N}$ refer to generalized point sets where each $\mathbf{x}_i$ lies in some space. An ordinary point set has $\mathbf{x}_i = x_i \in \mathbb{R}^d$. An oriented point set has $\mathbf{x}_i = (x_i, \hat{x}_i)$ where each $\hat{x}_i$ is a unit vector. Assuming a smooth underlying surface or contour, the action of a transformation $T : \mathbb{R}^d \rightarrow \mathbb{R}^d$ on orientation vector $\hat{x}$ at point x is $T \star \hat{x} = (J_T(x)^{-\mathrm{T}}\hat{x})/\|J_T(x)^{-\mathrm{T}}\hat{x}\|$, where $J_T(x)$ is the Jacobian at x. Note that this operation $\star$ respects composition in the sense that $(T_1 \circ T_2) \star \hat{x} = T_1 \star (T_2 \star \hat{x})$. For notational convenience, we will write $T(x, \hat{x}) = (T(x), T \star \hat{x})$, and let $T(\mathbf{X}) = \{T(\mathbf{x}) \mid \mathbf{x} \in \mathbf{X}\}$.

Suppose $\mathbf{X}$ and $\mathbf{Y}$ are generalized point sets of size N and M respectively, and let T be a transformation. Applying T to $\mathbf{Y}$, construct the mixture model

$$p(\mathbf{x}; T, \Theta) = \omega\eta(\mathbf{x}) + (1 - \omega)\sum_{j=1}^{M}\frac{1}{M}f(\mathbf{x}; T(\mathbf{y}_j), \theta_j),$$

where $\Theta = \{\theta_j\}$ is a set of additional parameters for distributions $f(\mathbf{x}_i; T(\mathbf{y}_j), \theta_j)$ centred on $T(\mathbf{y}_j)$, $\eta(\mathbf{x})$ is a uniform distribution to capture outliers, and ω is a hyperparameter indicating the expected proportion of outliers. For target point set $\mathbf{X}$, the log-likelihood is then $\log p(\mathbf{X}; T, \Theta) = \sum_{i=1}^{N} \log p(\mathbf{x}_i; T, \Theta)$. Introducing latent indicator variables z_i, where $z_i = j$ indicates $\mathbf{x}_i$ was sampled from the jth term in the mixture (or 0 for outliers), the posterior matrix computed during the E-step is

$$p_{ji} = p(z_i = j \mid \mathbf{x}_i; T^t, \theta_j^t) = \frac{(1 - \omega)f(\mathbf{x}_i; T^t(y_j), \theta_j^t)}{Mp(\mathbf{x}_i; T^t, \Theta^t)}, \tag{1}$$

where $\{T^t, \Theta^t\}$ are the parameters estimated at the previous iteration.

For ordinary point sets, Gaussians are a common choice; $f(x_i; T(\mathbf{y}_j), \theta_j) = \mathcal{N}(x_i; T(y_j), \sigma^2)$, where σ^2 is a single isotropic variance shared by all terms of the mixture, as in coherent point drift (CPD) [19]. Similar to other existing work [18], the mixture model can be extended for oriented points by choosing $f(\mathbf{x}_i; T(\mathbf{y}_j), \theta_j) = \mathcal{N}(x_i; T(y_j), \sigma^2)\mathcal{V}(\hat{x}_i; T \star \hat{y}_j, \kappa)$ where $\mathcal{V}(\hat{x}_i; T \star \hat{y}_j) = C_d(\kappa)\exp(\kappa\hat{x}_i^{\mathrm{T}}(T \star \hat{y}_j))$ are von Mises Fisher distributions with a single shared concentration parameter κ, and $C_d(\kappa)$ is a normalization coefficient whose form depends on d. These models respectively result in negative expected complete-data log-likelihood functions

$$Q(T, \sigma^2) = \sum_{i,j} \frac{p_{ji}}{2\sigma^2} \|x_i - T(y_j)\|^2 + \sum_{i,j} p_{ji} \log(\sigma^2) + R(T),$$

and

$$Q(T, \sigma^2, \kappa) = \sum_{i,j} p_{ji} \left(\frac{\|x_i - T(y_j)\|^2}{2\sigma^2} - \kappa(\hat{x}_i^T(T \star \hat{y}_j)) \right)$$

$$+ \sum_{i,j} p_{ji} \log(\sigma^2) - \sum_{i,j} p_{ji} \log(C_d(\kappa)) + R(T),$$

where $R(T)$ is an optional regularization on the transformation T.

For a true EM algorithm, the M-step consists of the optimization of $Q(T, \Theta)$ with respect to all parameters simultaneously. More common for point set registration is the expectation/conditional-maximization (ECM) algorithm [17], as used in CPD, breaking the M-step into CM-steps, wherein subsets of the parameters may be considered for optimization separately. A typical case would be to optimize Q first with respect to the transformation T, then with respect to the remaining statistical parameters. ECM algorithms retain the convergence properties of EM algorithms—namely, they will converge to a stationary point of the likelihood function. In contrast, generalized expectation-maximization (GEM) algorithms relax the requirements on the M-step such that Q need only be reduced, rather than minimized [7]. GEM algorithms can only be guaranteed to monotonically improve the overall likelihood, but have in spite of this fact been successfully employed for point set registration in recent years, especially when the chosen formulation results in a particularly complicated M-step [4,8,22].

We will break the present M-step into separate CM-steps for the transformation and distribution parameters. For points, these are

$$T' = \arg\min_{T} \sum_{j=1}^{M} w_j \|v_j - T(y_j)\|^2 + R(T); \tag{2}$$

$$\sigma^2 = \frac{\sum_{i,j} p_{ji} \|x_i - T(y_j)\|^2}{\sum_{i,j} p_{ji}}; \tag{3}$$

where $w_j = \sum_i p_{ji}$, $v_j = (\sum_i p_{ji} x_i)/w_j$ are the virtual observations [11]. For oriented points, the transformation CM-step can be written as

$$T' = \arg\min_{T} \sum_{j} w_j \|v_j - T(y_j)\|^2 - \sum_{j} \hat{w}_j \hat{v}_j \cdot (T \star \hat{y}_j) + R(T), \tag{4}$$

defining $\hat{w}_j = \sum_i \hat{p}_{ji}$, and $\hat{v}_j = (\sum_i \hat{p}_{ji} \hat{x}_i)/\hat{w}_j$ as an extension to the concept of virtual observations. The distribution CM-step can update σ^2 exactly as in Eq. (3), and κ by noting that $C_d(\kappa)$ can be expressed for arbitrary dimension in terms of the modified Bessel functions of the first kind I_ν, with a simpler analytic form for $d = 3$:

$$C_d(\kappa) = \frac{\kappa^{d/2-1}}{(2\pi)^{d/2} I_{d/2-1}(\kappa)};$$

$$C_3(\kappa) = \frac{\kappa}{4\pi \sinh \kappa};$$

due to the identity $\sinh x = \sqrt{\frac{\pi x}{2}} I_{1/2}(x)$. Given this, the CM-step for κ can be solved with a simple iterative algorithm based on the work of Sra [25]:

$$\bar{R} = \frac{\sum_{i,j} p_{ji} (T \star \hat{y}_j)^T \hat{x}_i}{\sum_{i,j} p_{ji}}; \quad A_d(x) = I_{d/2}(x)/I_{d/2-1}(x) \tag{5}$$
$$\kappa_0 = \frac{\bar{R}(d - \bar{R}^2)}{1 - \bar{R}^2}; \quad \kappa_{t+1} = \kappa_t - \frac{A_d(\kappa_t) - \bar{R}}{1 - A_d(\kappa_t)^2 - (d-1)A_d(\kappa_t)/\kappa_t}.$$

For the special case of $d = 3$, a rather simpler fixed point iteration can be derived [24]: $\kappa_{t+1}^{-1} \leftarrow \coth \kappa_t - \bar{R}$.

The form of the optimization problems to be solved during the transformation CM-step—Eq. (2) and Eq. (4) for points and oriented points respectively—define what we will call the transform update equation (TUE).

3 The Proposed Algorithm

Suppose we have nested transformation $T = T_1 \circ T_2 \circ \cdots \circ T_k$, and define

$$T_{i,j} = \begin{cases} T_i \circ \cdots \circ T_j, & \text{if } i < j, \\ T_i, & \text{if } i = j, \\ \text{Id}, & \text{if } i > j. \end{cases}$$

Assume that for each constituent transformation T_i there is a known equation or algorithm TUE_{T_i} where $\hat{T}_i = \text{TUE}_{T_i}(\mathbf{V} = \{\mathbf{v}_j\}, \mathbf{Y} = \{\mathbf{y}_j\}, W = \{w_j, \hat{w}_j\})$ solves the TUE as defined above for the single transformation T_i.[1] In this section we give the core contribution of this work: we will define an algorithm which can update the arbitrary composed transformation T during the CM-step, relying only on the individual solutions TUE_{T_i}.

Ideally we would define a procedure that would effectively act as TUE_T—that is, the solution of the TUE for the composed transformation. This suggests a function signature $T' = \text{UPDATETRANSFORMATION}(\mathbf{V}, \mathbf{Y}, W, T)$.

If the input transform T consists of a single atomic transformation T_1—that is, $k = 1$—then the procedure should simply apply the known solution TUE_{T_1}. If $k > 1$ the procedure could update the parameters of the outermost transformation T_1, and recursively apply itself to $T_{2,k}$, though these steps need not necessarily be taken in that order. The difficulty of designing such a procedure is that because the output of $T_{2,k}$ is then passed to T_1 to get the final output

---

[1] W may or may not contain $\{\hat{w}_i\}$ depending on if we are using oriented points.

of T, in general the optimal parameters of $T_{2,k}$ will depend on the parameters of T_1. Such a dependence would violate the goal of modularity gained by having the update rely only on the individual solutions TUE_{T_i}.

In certain special cases it will be possible to account for the outermost transformation in the recursive call to UPDATETRANSFORMATION by modifying the inputs—namely, through the target point set and weights. In the remainder of cases, by requiring only that the algorithm be a GEM algorithm, a simple heuristic approach to update the transformations can be used, which, combined with a fallback strategy is sufficient to ensure the overall reduction of the negative expected complete data log-likelihood during the resulting M-step.

3.1   The Special Case

Suppose T_1 is invertible and there is a deterministic mapping $\mathcal{I}_{T_1}$ of weights $W \mapsto W'$ given T_1 such that for all j and for any value of $T_{2,k}$

$$w_j \|v_j - T_1 \circ T_{2,k}(y_j)\|^2 = w_j' \|T_1^{-1}(v_j) - T_{2,k}(y_j)\|^2,$$

and, if applicable

$$\hat{w}_j \hat{v}_j^T ((T_1 \circ T_{2,k}) \star \hat{y}_j) = \hat{w}_j' (T_1^{-1} \star \hat{v}_j)^T (T_{2,k} \star \hat{y}_j).$$

Define transformations with such a mapping as TUE-invertible. Intuitively, TUE-invertibility means that the TUE for the overall transform T, target $\mathbf{V}$, and model $\mathbf{Y}$, with T_1 held constant, can be translated into an equivalent TUE for transform $T_{2,k}$, target $T_1^{-1}(\mathbf{V})$, and model $\mathbf{Y}$. Then UPDATETRANSFORMATION can correctly account for T_1 when updating $T_{2,k}$ as follows:

$$T_1' = \text{TUE}_{T_1}(\mathbf{V}, T_2(\mathbf{Y}), W);$$
$$T_{2,k}' = \text{UPDATETRANSFORMATION}((T_1')^{-1}(\mathbf{V}), \mathbf{Y}, \mathcal{I}_{T_1'}(W), T_{2,k}).$$

3.2   A General Heuristic Approach

If T_1 is not TUE-invertible, a simple heuristic approach is to update $T_{2,k}$ ignoring T_1, and then pass the point set $T_{2,k}'(\mathbf{Y})$ to the update for T_1:

$$T_{2,k}' = \text{UPDATETRANSFORMATION}(\mathbf{V}, \mathbf{Y}, W, T_{2,k});$$
$$T_1' = \text{TUE}_{T_1}(\mathbf{V}, T_{2,k}'(\mathbf{Y}), W).$$

This generalizes the heuristic used for nested affine transformations in the ultrasound processing from our previous work [22]. Though the heuristic ignores T_1 when updating $T_{2,k}$, it is still partly accounted for through the weights computed during the E-step.

3.3 The Complete Algorithm and Fallback Strategy

A pseudocode specification of UPDATETRANSFORMATION combining the two strategies described above is given in the bottom of Algorithm 1. However, when applying the heuristic strategy, there is still no guarantee that UPDATETRANS-FORMATION will improve the expected complete-data log-likelihood, which is necessary to ensure the resulting algorithm is a GEM algorithm [7]. To achieve this, a simple fallback strategy can be used as specified in GEMUPDATETRANSFOR-MATION in Algorithm 1: prior to updating the transformations, compute $Q(T, \Theta)$. Then, if after updating all parameters, $Q(T', \Theta) > Q(T, \Theta)$, the parameters of T_k can be reverted to their initial values, and $T_{1,k-1}$ updated instead. Similarly, if the complete-data log-likelihood is still worse, T_{k-2} can be fixed to its initial value, and so on. In the worst case, only T_1 is updated, and $Q(T', \Theta) \leq Q(T, \Theta)$ is guaranteed by definition of TUE_{T_1}.

Algorithm 1. A complete pseudocode specification of the transformation CM-step. $\Theta = \{\sigma^2\}$ for points, and $\Theta = \{\sigma^2, \kappa\}$ for oriented points.

1: **procedure** GEMUPDATETRANSFORMATION($\mathbf{V}, \mathbf{Y}, W, T, \Theta$)
2:     $E \leftarrow Q(T, \Theta)$, $E' \leftarrow \infty$, $s \leftarrow k$
3:     **while** $E' \geq E$ **do**
4:         $T'_{1,s} \leftarrow$ UPDATETRANSFORMATION($\mathbf{V}, T_{s+1,k}(\mathbf{Y}), W, T_{1,s}$)
5:         $E' \leftarrow Q(T', \Theta)$, $s \leftarrow s - 1$
6:     **end while**
7:     **return** T'
8: **end procedure**
9: **procedure** UPDATETRANSFORMATION($\mathbf{V}, \mathbf{Y}, W, T$)
10:     $\ell \leftarrow$ number of transformations in T
11:     **if** $\ell = 1$ **then**
12:         **return** $\text{TUE}_{T_1}(\mathbf{V}, \mathbf{Y}, W)$
13:     **end if**
14:     **if** T_1 is TUE-invertible **then**
15:         $T'_1 \leftarrow \text{TUE}_{T_1}(\mathbf{V}, T_{2,\ell}(\mathbf{Y}), W)$
16:         $T'_{2,\ell} = $ UPDATETRANSFORMATION($(T'_1)^{-1}(\mathbf{V}), \mathbf{Y}, \mathcal{I}_{T'_1}(W), T_{2,\ell}$)
17:     **else**
18:         $T'_{2,\ell} \leftarrow$ UPDATETRANSFORMATION($\mathbf{V}, \mathbf{Y}, W, T_{2,\ell}$)
19:         $T'_1 \leftarrow \text{TUE}_{T_1}(\mathbf{V}, T'_{2,\ell}(\mathbf{Y}), W)$
20:     **end if**
21:     **return** $T'_1 \circ T'_{2,\ell}$
22: **end procedure**

To register $\mathbf{X}$ and $\mathbf{Y}$, simply iterate the following until convergence:

- E-step: compute the posteriors as in Eq. (1);
- Transformation CM-step: Update T with Algorithm 1; and
- Distribution CM-step: Update σ^2 with Eq. (3), and κ with Eq. (5) if applicable.

3.4   Basic Transformations

The basic transformations used throughout the experiments are rigid, affine, and nonrigid Gaussian radial basis functions (GRBFs), as well as rotations about a fixed point. Of these, only the rigid/rotations are TUE-invertible. Any of these can be made to apply to only a subset of the point set, because the solution to the TUE in such a case will simply be the solution to the TUE with only those transformed points and their corresponding target points being input. This allows the construction of relatively complex articulated transformations.

For point sets, solutions for all of these basic transformations were given in CPD [19]. A solution for rigid registration with oriented points has also been published [18]. For oriented points with affine or GRBF transformations we used a gradient-based optimization as in our previous work [22].

4   Experiments

Experiments were conducted on the models shown in Fig. 1: a 3D point set depicting a bug, and a 2D point set depicting a left cardiac ventricle model. The bug was constructed in terms of simple shapes, consisting of 126 oriented points in total. The left ventricle was based on a manual outline in a 2D slice of a 3D ultrasound image from the STACOM 2011 MICCAI cardiac motion tracking challenge [27]. Both were centered at the origin, and scaled such that root mean square (RMS) distance of the points from the origin was 1.

 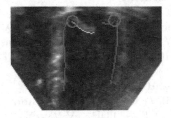

Fig. 1. Initial bug and left ventricle models. (Color figure online)

The transformation of the bug was described by a global rigid transformation R, composed with six additional rotations L_i—one per leg—each rotating about the meeting point of the leg and body, giving $T = R \circ L_1 \circ L_2 \circ L_3 \circ L_4 \circ L_5 \circ L_6$. The leg rotations L_i were regularized with $R(L_i) = \lambda_L \|L_i - I\|_F^2$ where $\lambda_L = 1$ was a fixed regularization weight, and $\|\cdot\|_F$ is the Frobenius norm.

The transformation of the left ventricle was composed of $T = G \circ R_1 \circ R_2 \circ A$ where A was an affine transformation, R_1 and R_2 were rotations of points outlining either side of the mitral valve about the point nearest to the wall, and G was a nonrigid GRBF transformation with width parameter $\beta = 0.7$. The subsets to which R_1 and R_2 applied are shown in green and purple in Fig. 1, while the centers of rotation are circled in orange. The rotations were meant to capture the opening and closing of the mitral valve, while the affine and Gaussian radial basis function (GRBF) transformations were intended to capture the large-scale and small scale deformations that the ventricle might undergo, respectively. During registration, the only regularization was that of the GRBF, with weight $\lambda = 5$.

For both models, 25 target point sets were generated by applying random transformations to the model. These served as ground truth in experiments to examine the robustness to noise and to outliers. For the noise experiments, randomly chosen Gaussian distributed offsets were applied to the points of each target set at 10 different standard deviations from 0 up to 0.2. Noise was added to the orientations by choosing normal vectors distributed uniformly over the sphere until the angle in radians was less than the current noise level. For the outlier experiments, in addition to noise at level 0.02, uniformly placed outliers were added to the targets at 10 different proportions from 0 to 0.5. All outlier orientations were uniformly distributed over the sphere. The accuracy of the registration was evaluated in terms of the pairwise RMS error between the ground truth correspondences from model to target, before adding noise or outliers.

As a quantitative comparison between the performance of the algorithm with or without oriented points, a statistical decision procedure was used to select for each experiment, whether the registration with points or oriented points was preferred (possibly concluding neither in the case of insufficient evidence) [14, pp. 27–29]. This procedure essentially entails doing two one-sided Wilcoxin signed-rank tests at half the desired confidence level, testing for improvement in both directions. One of the two tests will be significant if and only if the equivalent two-sided test would be significant. That test which was significant indicates the preferred method. In all experiments described below, confidence level $\alpha = 0.001$ was used as the threshold for significance.

Table 1. Statistical tests comparing the registrations with points (P) and oriented points (OP). Tests significant at level $\alpha = 0.001$ are bold, and list the preferred method.

Noise	p-value				Outlier	p-value			
	Bug		Left ventricle			Bug		Left ventricle	
0.000	**1.2e-05**	(OP)	**3.6e-05**	(OP)	0.000	**1.2e-05**	(OP)	**3.2e-05**	(OP)
0.022	**1.2e-05**	(OP)	**2.9e-05**	(OP)	0.056	**1.2e-05**	(OP)	**4.1e-05**	(OP)
0.044	**1.4e-05**	(OP)	**0.0004**	(OP)	0.111	**1.4e-05**	(OP)	**0.00013**	(OP)
0.067	**2.3e-05**	(OP)	0.0027		0.167	**1.2e-05**	(OP)	**0.00016**	(OP)
0.089	**2.3e-05**	(OP)	0.0025		0.222	**1.2e-05**	(OP)	**0.00049**	(OP)
0.111	**4.1e-05**	(OP)	0.0025		0.278	**1.2e-05**	(OP)	**0.00014**	(OP)
0.133	**9.0e-05**	(OP)	**0.00014**	(OP)	0.333	**1.2e-05**	(OP)	**4.1e-05**	(OP)
0.156	**2.5e-05**	(OP)	0.0035		0.389	**1.2e-05**	(OP)	**0.0003**	(OP)
0.178	**0.0001**	(OP)	0.0013		0.444	**1.2e-05**	(OP)	**0.00045**	(OP)
0.200	**8.1e-05**	(OP)	0.0021		0.500	**2.0e-05**	(OP)	**3.6e-05**	(OP)
Pooled	**9.8e-36**	(OP)	**1.6e-27**	(OP)	**Pooled**	**1.9e-42**	(OP)	**8.2e-35**	(OP)

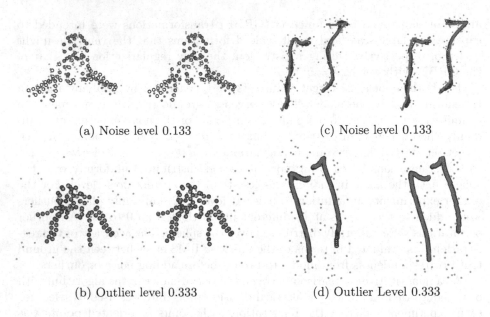

(a) Noise level 0.133   (c) Noise level 0.133

(b) Outlier level 0.333   (d) Outlier Level 0.333

Fig. 2. Sample registrations done with points (left) and oriented points (right). The transformed model is blue, the ground truth target is red, and the actual target is gray. (Color figure online)

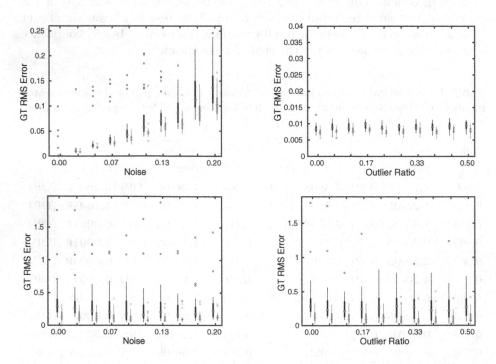

Fig. 3. The RMS errors for points (blue) and oriented points (green), for the bug (top) and left ventricle (bottom) noise (left) and outlier (right) experiments. (Color figure online)

The results of the statistical decision procedure for all experiments are summarized in Table 1. All conclusive experiments were in favour of oriented points. Figure 2 shows some sample registrations comparing the results using points and oriented points. For each noise/outlier level shown, the case nearest to the median RMS error for the oriented point experiments was chosen. The RMS error distributions for all experiments are summarized in box plots in Fig. 3.

Table 2. The fractions of iterations in which each subset of the left ventricle transformation was updated across all experiments.

Updated	1	1–2	1–3	1–4
Points	0.17	0.03	0.05	0.75
Oriented points	0.06	0.14	0.02	0.78

Table 2 summarizes how often the fallback strategy in GEMUPDATETRANS-FORMATION was needed in the left ventricle experiments. Results for the bug are not included because all of the bug transformations are TUE-invertible, and therefore the heuristic update is never applied. Finally, the median timing for a single registration across all noise and outlier experiments for each of the four configurations (bug and left ventricle, each with points or oriented points) is given in Table 3.

Table 3. Median time in seconds to register with GEM-basic. All timings were obtained on an older computer running Ubuntu 16.04 with a 4-core Intel Core i7-2600K CPU and 12 GiB memory.

Bug		Left Ventricle	
Points	Oriented points	Points	Oriented points
0.06	0.07	0.38	**133.62**

5   Discussion

The value of a generic algorithm allowing the modular composition of transformations can be seen relative to other EM-based registration algorithms making use of composed transformations. For example, composed models of transformations have been used for the registration of models of both the spine and wrist [1,23]. Both models consisted of composed statistical models of shape and pose, for which a gradient based optimization was used during the transformation CM-step. The gradients with respect to the pose and shape parameters were interdependent, and thus had to be derived for the specific chain of transformations in use. Likewise, Khallaghi et al. [12] composed a similarity transformation with a statistical shape model and an additional biomechanical nonrigid transformation for a prostate registration problem. The transformation CM-step was broken into a sequence of three CM-steps, one per transformation; these updates were interdependent, each taking into account the current parameters controlling all of the other transformations. In comparison, our algorithm can be applied to any composed transformation so long as each constituent transform has a known solution to the TUE. Furthermore, our approach was easily formulated for both points and oriented points, whereas the aforementioned methods were all restricted to points.

In some cases, including that of the bug model examined here, the proposed algorithm will be a true ECM algorithm. In particular, this will be the case for $T = T_1 \circ \cdots \circ T_k$, where T_i are TUE-invertible for all $i < k$. This is because TUE-invertible transformations can be correctly accounted for by those transformations that come after as discussed in Subsect. 3.1.

In all other cases, the proposed algorithm relies on a heuristic approach to update the component transformations. In order to guarantee that the algorithm is a GEM algorithm, it must have the fallback strategy of updating successively smaller subsets of the full transformation until the expected complete-data log-likelihood is improved. From Table 2, in the case of the points experiments, all four transformations were updated 75% of the time. For oriented points, all four transformations were updated 78% of the time. Therefore, it appears that the heuristic was effective in the left ventricle experiments the majority of the time, but the fallback strategy was still important in maintaining GEM status.

The accuracy of the registration was evaluated in terms of the pairwise RMS error between the ground truth correspondences between model and target, before the addition of noise and/or outliers. Although the true transformation parameters for each target were also known, we chose to instead evaluate the registrations spatially because some choices of transformation spaces may be such that a single transformation has more than one possible parameterization. As a simple hypothetical example, a composed rigid and affine transformation, wherein the affine part is the identity, could be exactly reproduced with different parameters by setting the rigid parameters as the identity, and encoding the rotation matrix into the affine transformation instead. Even for the left ventricle, suppose the transformation T was chosen with all component transformations as the identity, except for the affine transformation A. A perfect registration could in principle be achieved between all N points in the model using only the parameters of the GRBF G by choosing the weights that interpolate between the N samples mapping y_j to v_j.

Considering the results in Table 1 it appears, broadly speaking, that the inclusion of oriented points improved the accuracy of the algorithm as measured by this metric. Looking at the qualitative results for the bug in Fig. 2, the results for points and oriented points are difficult to distinguish. This is likely due to the fact that the median errors are mostly very small in all experiments, so the results, despite being statistically significantly different, are not particularly consequential in terms of true registration quality. In contrast, the qualitative results for the left ventricle in Fig. 2 do seem to show a distinct improvement when registering with oriented points, at least in some cases, despite the statistical tests being less decisively in favour of oriented points over points.

Looking in particular at Fig. 2(c), one might note the improved registration of the mitral valve. One contributing factor to this effect may be that the points on the left part of the mitral valve and the wall are relatively close together in position, but nearly opposite in orientation. Accordingly, when registering without orientation information, those points are harder to distinguish. We observed a similar effect in the cardiac ultrasound registration algorithm presented in our previous work [22], where the inclusion of orientations helped prevent points on either side of the septum from drifting to the opposite side. However, whatever accuracy gains may be had by including orientations does come at a computational cost associated with the solving of the rather more complex TUE for oriented points compared to that for points.

The median timings in Table 3 give an idea of the variability of this cost. First, it is notable that for both points and oriented points the left ventricle registration is considerably more costly than that of the bug. The GRBF transformation, with its large weight matrix, requires more computation than the rigid transformations employed by the bug model. The most striking result, however, is the drastically larger median time for oriented points in the left ventricle. This discrepancy is due to the use of a gradient based solver for the oriented point TUE for the GRBF transformation, which in this case had hundreds of parameters to optimize. Better algorithms for solving the oriented point TUE for these cases would therefore be a useful contribution, could they be found.

While the mathematical framing in Sect. 2 is given for a particular choice of mixture models, it should be noted that the proposed algorithm really only relies on the forms of the transformation CM-steps in Eq. (2) and Eq. (4). This form is not necessarily limited to the particular mixture models used here. For example, a TUE of the form of Eq. (2) can also be used for models using Student's t-distributions [29], or heteroscedastic covariances [10]. This suggests that the algorithm may be generalizable to a broader class of mixture models.

It remains an open question how the results presented here will generalize to other applications. Validation of registration algorithms is in general complicated by the question of generalization to other datasets, but the proposed algorithm is also generic in terms of transformation. Interestingly, this can be seen as both a strength and a weakness for the purposes of generalization. For a registration algorithm using a fixed, general model of transformation like non-rigid CPD [19], the question is if the transformation model will generalize well to different applications. In contrast, our algorithm is designed to facilitate the use of transformations tailored to a given application. However, this flexibility comes at the expense of making it difficult to determine how well the algorithm will perform on the limitless array of composed transformations to which it could, in principle, be applied. Furthermore, the focus on composed transformations makes the fair comparison to existing state of the art approaches difficult.

6   Conclusion

We proposed a GEM algorithm to allow registration of point or oriented point sets through arbitrarily composed sequences of transformations in a modular way, and presented experiments on two models with varying combinations of transformations. Our results show the algorithm to be effective across a wide range of noise and outliers, and consistently favour the inclusion of orientation information, with the necessary caveat that the generic nature of the algorithm makes it difficult to generalize the results. Further studies applying the method in other application contexts would therefore be a valuable contribution. Other opportunities for future work include the generalization of the algorithm to other types of mixture models, and improved algorithms for solving the TUE for affine or GRBF transformations with oriented points. Ultimately, it is our hope that the work presented here can encourage and simplify the use of application-specific models of deformation to improve registration in the medical imaging context.

Acknowledgements. This work was supported in part by the Natural Sciences and Engineering Research Council of Canada under Grant RGPIN-2018-04430.

References

1. Anas, E.M.A., et al.: Automatic segmentation of wrist bones in CT using a statistical wrist shape+pose model. IEEE Trans. Image Process. **35**(8), 1789–1801 (2016)
2. Anguelov, D., Srinivasan, P., Koller, D., Thrun, S., Rodgers, J., Davis, J.: SCAPE: shape completion and animation of people. In: SIGGRAPH 2005: ACM SIGGRAPH 2005 Papers, pp. 408–416 (2005)
3. Baka, N., Metz, C.T., Schultz, C.J., van Geuns, R., Niessen, W.J., van Walsum, T.: Oriented Gaussian mixture models for nonrigid 2D/3D coronary artery registration. IEEE Trans. Med. Imaging **33**(5), 1023–1034 (2014)
4. Bernard, F., et al.: Shape-aware surface reconstruction from sparse 3D point-clouds. Med. Image Anal. **38**, 77–89 (2017)
5. Billings, S., Taylor, R.: Generalized iterative most likely oriented-point (G-IMLOP) registration. Int. J. Comput. Assist. Radiol. Surg. **10**(8), 1213–1226 (2015). https://doi.org/10.1007/s11548-015-1221-2
6. Cootes, T.F., Taylor, C.J., Cooper, D.H., Graham, J.: Active shape models–their training and application. Comput. Vis. Image Underst. **61**(1), 38–59 (1995)
7. Dempster, A.P., Laird, N.M., Rubin, D.B.: Maximum likelihood from incomplete data via the EM algorithm. J. R. Stat. Soc. Ser. B Methodol. **39**(1), 1–38 (1977)
8. Eckart, B.: Compact generative models of point cloud data for 3D perception. Ph.D. thesis, Carnegie Mellon University Pittsburgh (2017)
9. Erdt, M., Steger, S., Kirschner, M., Wesarg, S.: Fast automatic liver segmentation combining learned shape priors with observed shape deviation. In: Proceedings of the IEEE Symposium on Computer-Based Medical Systems, pp. 249–254 (2010)
10. Evangelidis, G.D., Horaud, R.: Joint alignment of multiple point sets with batch and incremental expectation-maximization. IEEE Trans. Pattern Anal. Mach. Intell. **40**(6), 1397–1410 (2018)

11. Horaud, R., Forbes, F., Yguel, M., Dewaele, G., Zhang, J.: Rigid and articulated point registration with expectation conditional maximization. IEEE Trans. Pattern Anal. Mach. Intell. **33**(3), 587–602 (2011)
12. Khallaghi, S., et al.: Statistical biomechanical surface registration: Application to MR-TRUS fusion for prostate interventions. IEEE Trans. Med. Imaging **34**(12), 2535–2549 (2015)
13. Khallaghi, S., et al.: Biomechanically constrained surface registration: application to MR-TRUS fusion for prostate interventions. IEEE Trans. Med. Imaging **34**(11), 2404–2414 (2015)
14. Lehmann, E.L.: Nonparametrics: Statistical Methods Based on Ranks, Revised 1st edn. Springer, New York (2006)
15. Loper, M., Mahmood, N., Romero, J., Pons-Moll, G., Black, M.J.: SMPL: a skinned multi-person linear model. ACM Trans. Graph. **34**(6) (2015)
16. Maiseli, B., Gu, Y., Gao, H.: Recent developments and trends in point set registration methods. J. Vis. Commun. Image Represent. **46**(C), 95–106 (2017)
17. Meng, X.L., Rubin, D.B.: Maximum likelihood estimation via the ECM algorithm: a general framework. Biometrika **80**(2), 267–278 (1993)
18. Min, Z., Wang, J., Meng, M.Q.: Robust generalized point cloud registration with orientational data based on expectation maximization. IEEE Trans. Autom. Sci. Eng. **17**(1), 207–221 (2020)
19. Myronenko, A., Song, X.: Point set registration: coherent point drift. IEEE Trans. Pattern Anal. Mach. Intell. **32**(12), 2262–2275 (2010)
20. Okada, T., Linguraru, M.G., Hori, M., Summers, R.M., Tomiyama, N., Sato, Y.: Abdominal multi-organ segmentation from CT images using conditional shape-location and unsupervised intensity priors. Med. Image Anal. **26**(1), 1–18 (2015)
21. Oliveira, F.P., Tavares, J.M.R.: Medical image registration: a review. Comput. Methods Biomech. Biomed. Eng. **17**(2), 73–93 (2014)
22. Peoples, J.J., Bisleri, G., Ellis, R.E.: Deformable multimodal registration for navigation in beating-heart cardiac surgery. Int. J. Comput. Assist. Radiol. Surg. **14**(6), 955–966 (2019). https://doi.org/10.1007/s11548-019-01932-2
23. Rasoulian, A., Rohling, R., Abolmaesumi, P.: Lumbar spine segmentation using a statistical multi-vertebrae anatomical shape+pose model. IEEE Trans. Med. Imaging **32**(10), 1890–1900 (2013)
24. Ravikumar, N., Gooya, A., Frangi, A.F., Taylor, Z.A.: Generalised coherent point drift for group-wise registration of multi-dimensional point sets. In: Descoteaux, M., Maier-Hein, L., Franz, A., Jannin, P., Collins, D.L., Duchesne, S. (eds.) MICCAI 2017. LNCS, vol. 10433, pp. 309–316. Springer, Cham (2017). https://doi.org/10.1007/978-3-319-66182-7_36
25. Sra, S.: A short note on parameter approximation for von Mises-Fisher distributions: and a fast implementation of $I_s(x)$. Comput. Stat. **27**(1), 177–190 (2012)
26. Tam, G.K.L., et al.: Registration of 3D point clouds and meshes: a survey from rigid to nonrigid. IEEE Trans. Vis. Comput. Graph. **19**(7), 1199–1217 (2013)
27. Tobon-Gomez, C., et al.: Benchmarking framework for myocardial tracking and deformation algorithms: an open access database. Med. Image Anal. **17**(6), 632–648 (2013)
28. Yokota, F., et al.: Automated CT segmentation of diseased hip using hierarchical and conditional statistical shape models. In: Mori, K., Sakuma, I., Sato, Y., Barillot, C., Navab, N. (eds.) MICCAI 2013. LNCS, vol. 8150, pp. 190–197. Springer, Heidelberg (2013). https://doi.org/10.1007/978-3-642-40763-5_24
29. Zhou, Z., Zheng, J., Dai, Y., Zhou, Z., Chen, S.: Robust non-rigid point set registration using Student's-t mixture model. PLoS ONE **9**(3), 1–11 (2014)

Uncertainty Reduction in Contour-Based 3D/2D Registration of Bone Surfaces

Xolisile O. Thusini[1(✉)], Cornelius J. F. Reyneke[1], Jonathan Aellen[3],
Andreas Forster[3], Jean-Rassaire Fouefack[1,2], Nicolas H. Nbonsou Tegang[1],
Thomas Vetter[3], Tania S. Douglas[1], and Tinashe E. M. Mutsvangwa[1,3]

[1] Division of Biomedical Engineering, University of Cape Town,
Cape Town, South Africa
thsxol001@myuct.ac.za

[2] Department of Image and Information Processing, IMT-Atlantique, Brest, France

[3] Department of Mathematics and Computer Science, University of Basel, Basel,
Switzerland

Abstract. The reconstruction of $3D$ bone shape from $2D$ X-ray contours is an ill-posed problem. For medical applications, it is important to estimate the uncertainty of the reconstructions. While traditional optimisation methods produce a single point-estimate, we frame the problem as Bayesian inference. We apply a Monte Carlo sampling based non-rigid $3D$ to $2D$ registration recovering the posterior distribution of plausible reconstructions. This provides insight into the uncertainty of the inferred $3D$ reconstruction. As an application, we demonstrate the use of the method in selecting X-ray viewing conditions in order to maximise accuracy while minimising reconstruction uncertainty. We evaluated reconstruction accuracy and variance for the femur bone from bi-planar views.

Keywords: 3D reconstruction · Statistical shape model · $3D/2D$ registration · Bayesian inference · MCMC sampling · Optimal viewing-angle selection

1 Introduction

Non-rigid three-dimensional-to-two-dimensional $(3D/2D)$ registration may be used to obtain a patient-specific $3D$ surface reconstruction of a bone from a limited number of $2D$ patient X-ray images. Such a reconstruction is performed by iteratively adjusting the parameters of a $3D$ statistical shape model (SSM), based on the information inferred from the X-ray views. These surface reconstructions can then be used to aid in clinical tasks such as surgical planning and post-operative evaluation while avoiding the higher costs or relatively greater patient irradiation associated with some $3D$ modalities (computed tomography). The main clinical application area of such reconstructions is within orthopaedics. In order to reduce exposure to ionising radiation, and reduce computation time, the number of X-ray views is often limited to one or two (Hurvitz and Joskowicz

© Springer Nature Switzerland AG 2020
M. Reuter et al. (Eds.): ShapeMI 2020, LNCS 12474, pp. 18–29, 2020.
https://doi.org/10.1007/978-3-030-61056-2_2

2008). The viewing angle(s) are selected, either by adjusting the patient or by adjusting the C-arm of the X-ray machine (if one is available). In contour-based reconstruction, the chosen viewing angle(s), together with the type of bone, have an influence on the points that can reliably be identified, and thus also influence the uncertainty of the estimated patient-specific $3D$ bone surface. A decision, therefore, has to be made about which X-ray view(s) to use as many $3D$ shapes can account for the same X-ray image projection. Additionally, since different types of symmetry exist for different types of bone (femur vs. pelvis), the viewing angles that provide the least ambiguity - and maximise accuracy - are not always obvious (Suter et al. 2020). While the femur bone is relatively simple, and a 90° bi-planar viewing angle seems intuitive (since a maximum decorrelation of information between the two views in ensured) (Fleute and Lavallée 1999; Baka et al. 2011), bones such as the scapula or pelvis exhibit a profound increase in structural superposition for certain viewing angles. In their work, Sadowsky et al. (2007) opted to use a 90° and 45° separation for the pelvis (Sadowsky et al. 2007). It is evident that these methods rely mostly on user intuition, or heuristics, which lead to reconstruction error and/or inconsistent performance. Furthermore, maximum-point-estimate methods are employed, where the effect of different sources of uncertainty (such as $2D$ projection) are not investigated. Markov Chain Monte Carlo (MCMC) methods for image analysis are used to assess the uncertainty and have been reported to solve inference problems in object recognition, including the parsing of natural scene images, face analysis and segmentation (Zhu et al. 2000; Tu et al. 2005; Schönborn et al. 2017; and Egger 2017). The probabilistic nature of MCMC approaches allows reason about uncertainty of the image interpretation while simultaneously performing image registration.

Here we propose a framework for contour-based $3D$-from-$2D$ reconstruction. We frame the problem as model fitting based on Bayesian inference. We estimate the full posterior distribution of model instances given the image, and not only a single maximum-point-estimate as Sadowsky et al. (2007). The posterior has no closed form solution, and thus we resort to MCMC based approaches. We use the Metropolis-Hastings algorithm in combination with a given shape prior model considering minor pose variation. The inference is driven by a propose-and-verify architecture. A new sample is proposed from a proposal distribution given the current state. Then the sample is either accepted or rejected based on a acceptance criterion as originating from the posterior distribution over model parameters.

Since the exploration of the space with random proposals is slow at high dimensions and many samples are not independent, we propose the use of Hamiltonian Monte Carlo (HMC) as the second approach (Duane et al. 1987; Neal et al. 2011) to explore the search space more efficiently, while retaining accurate posteriors. We demonstrate how to use the inferred posterior distribution to compute the uncertainty of the reconstructed surface. Furthermore, we propose a method for automatic contour selection (on the $3D$ reference surface), capable of simulating the view-dependent contour, which can help determine how informative a given contour may be. By combining our algorithm with a contour generation method, we provide a means of quantifying uncertainty reduction for a particular X-ray view (or views). The efficacy of the framework is evaluated through a series of experiments using simulated images. Our contributions are as follows: 1) an MCMC-sampling method for $3D$-from-$2D$ bone reconstruction using the random walk Metropolis-Hastings method; 2) an improved sampling algorithm using the HMC method; 3) a method for estimating the uncertainty of a $3D$ reconstruction; and 4) an application of the proposed framework to determine optimal bi-planar X-ray viewing-angles.

2   Methods

2.1   Statistical Shape Models

We employ the method described by Lüthi et al. (2017) to train an SSM of the bone-of-interest. Building the SSM basically includes the extraction of the mean shape and a number of modes of variation across the individuals in the training dataset of typical shapes $\{\Gamma_n, n = 1, ..., N\}$ segmented from the CT scans of the population. The first step in computing the mean shape is to extract shape features from each training sample and align all the samples with respect to the reference shape using these features. The second step is the reduction of the training dataset into a small set of modes that best represent the observed shape variations. This is accomplished using principal component analysis (PCA) (Hotelling 1933). The PCA models only represent shapes that are in the linear span of the given training examples. The drawback of this specificity is that not all the target shapes can be fully represented in the model. To capture the complete PCA model space, a lot of training samples are needed which in reality is not a simple task. To overcome this, the work by Lüthi et al. (2017) extended the PCA models to include modeling of deformations using Gaussian process (GP). Hence the resulting model is the probability distribution defined on the deformations. Including the GP means that the deformations are modelled with Gaussian function $\mathcal{GP}(\mu, k)$, where μ and k are mean and covariance functions, respectively. In a Bayesian inferencing framework, the model is used as a prior distribution $P(\theta)$, where θ denotes the model parameters that are optimised during contour-fitting.

2.2   Synthetic Contour Generation, Projection and Detection

We demonstrate the validity of our approach in a simplified setting. To generate synthetic 2D images of contours, we produce a $2D$ silhouette from the 3D mesh. Our approach requires that we know the silhouette vertex points.

Our algorithm consists of two steps. First, we render a mesh as a binary image. Second, we find all vertices which account for the silhouette.

Let $M(\theta)$ be the mesh described by the model parameters θ. Next we render a binary occupancy image o_θ with the value $o_\theta(p) = 1$ when the object is visible at the pixel location p otherwise 0. Let N_p denote the set of adjacent pixels of pixel p, e denotes the eye vector, p_v the pixel location of vertex v in the binary image, S_v the set of adjacent surfaces of v with n_s denoting the normal of the surface $s \in S_v$. Then the collection of silhouette points C is defined as:

$$C = \{ \; v \mid v \in M(\theta), \; \exists s', s'' \in S_v, \; s' \neq s'' \text{ with } sgn(n_{s'} \cdot e) \neq sgn(n_{s''} \cdot e) \atop \text{and } \exists p' \in N_p \text{ with } o_\theta(p') = 0 \} \; . \tag{1}$$

The vertex set C in Eq. 1 contains all points which are on the silhouette and have adjacent surfaces which face in a different direction when projected onto the eye vector. Even though we generate a $2D$ silhouette image given a $3D$ shape, we still calculate distances using only the $2D$ silhouette.

This synthetic contour generation process is also used during the optimisation process in the model fitting.

2.3 Estimation of the Posterior Distribution

In order to estimate the posterior distribution $P(\theta|C)$ of the SSM, when provided with a set of pre-selected view-dependent $2D$ contours as observations, we replace a straight-forward optimisation approach with a MCMC-based inference algorithm. The inferred solution is a distribution, rather than a single maximum estimate.

Random Proposals. We adopt the MCMC approach of Schönborn et al. (2017), where the inference is driven by the proposed-verification architecture of the Metropolis-Hasting (MH) algorithm, as our first sampling approach. This algorithm draws random samples θ' from a proposal distribution $Q(\theta'|\theta)$ and transforms them into samples stemming from a target distribution $P(\theta|C)$ by accepting a proposed sample $\theta_{t+1} \to \theta'$ with the following probability:

$$a = \min\{\frac{P(\theta'|C)}{P(\theta|C)} \frac{Q(\theta|\theta')}{Q(\theta'|\theta)}, 1\}. \tag{2}$$

The algorithm may otherwise reject the sample and keep the current one $\theta_{t+1} \to \theta$. The function Q generates proposals θ' that are possible model parameter updates. Our implementation of the MCMC sampling-based registration algorithm integrates prior knowledge and contour observations using accept-reject filters. In this setting, we initialise the Markov chain as the model mean which later is updated after each accepted sample. Each accepted sample corresponds to a set of model parameters which provide an estimated bone surface. The resultant surface meshes are then used to represent the estimated posterior distribution for a specific set of contour points (corresponding to a specific X-ray view).

Guided Proposals. We use the HMC approach to efficiently explore the complex target distribution and obtain a more accurate variance estimator. The computational expense to create independent samples for a d-dimensional space generally scales with $d^{5/4}$, while it is d^2 for a random walk MH method. To define the Hamiltonian dynamics (Neal et al. 2011; Zoppo 2018), which create proposals which are nearly independent from the current state but still have a high acceptance rate, we extend a d-dimensional current state q with d-dimensional momentum variables p. The Hamiltonian function $H(q,p)$ is the sum of the potential energy $U(q)$ and the kinetic energy $K(p)$ formulated as:

$$H(q,p) = U(q) + K(p), \tag{3}$$

$$U(q) = -\log\left[P(q)l(q|C)\right] \text{ and } K(p) = p^T M^{-1} p/2.$$

Here the quadratic form of the kinetic energy is used to model the Gaussian targets. The mass matrix M is a free parameter for linear transformation. We use our estimate of the covariance matrix for M. The posterior distribution takes the role of the position variable, q which is expressed using the potential energy with $P(q)$ being the prior, and $l(q|C)$ is the likelihood.

Algorithm 1 explains how Eq. 3 is approximated for a given set of initial conditions, using a step-size ϵ, and number of leapfrog steps L. The dynamics are randomly perturbed by the momentum values drawn from the normal distribution to ensure that the space is explored at each sampling iteration.

Algorithm 1: Hamiltonean Monte Carlo.

HMC = f(U,gradU,ϵ,L,current_q)
Initialise
q= current_q
M $\leftarrow$ diagonal scale matrix
p $\leftarrow$ N(0,M)
current_p=p
p = p - ϵ * gradU(q)/2; % ensure volume preservation and reversibility
for $i=1,...,L$ **do**
    q= q + ϵ * p
    if $i \mathrel{!}= L$ **then**
        | p = p - ϵ * gradU(q)
    end
end
p = p - ϵ * gradU(q)/2
p = -p
if $N(0,1) \leq exp\left[U(current\_q) - U(q)|\frac{||current\_p||^2}{2} - \frac{||p||^2}{2}\right]$ **then**
    | **return** q
end
else
    | **return** current_q
end

Contour Likelihoods. We compute the posterior of the shape reconstruction $P(\theta|C)$ from a prior $P(\theta)$ and a contour likelihood $l(\theta; C)$

$$P(\theta|C) \propto l(\theta; C)P(\theta). \tag{4}$$

The likelihood functions are important in the acceptance step of both MC methods. They measure the quality of the image explanation by comparing the target contours to the occluding contours of the current model instance θ, rendered onto the image domain with projection function

$$f_\theta \to C' = \{p'_1, p'_2, ..., p'_i, ...\}. \tag{5}$$

Our contour likelihood relies on soft correspondence, meaning for each point of the model proposed contour we search for the closest point in the jth target contour line

$$C_j = \{p^j_1, p^j_2, ..., p^j_i, ...\}. \tag{6}$$

To that end, the contours of the target and the current model instance are made invariant under $3D$ motion, so that only shape variation remains between the two. For this we find the rigid transformation between C' and C_j as:

$$h^j = \arg \min_{h \in SE(3)} \| \overrightarrow{h(\{p^j_1, p^j_2, ..., p^j_i, ...\})} - \overrightarrow{\{f_\theta(p'_1), f_\theta(p'_2), ..., f_\theta(p'_i), ...\}} \|^2. \tag{7}$$

Our focus here is not on investigating the suitable likelihood function, thus we use the common assumption that the observed contour points are independent and are affected by additive Gaussian noise. Hence our likelihood rates the distances between closest points of the generated and the target contour after rigidly aligning them using a Gaussian noise model. We therefore define the likelihood as

$$l_j(\theta|C_j, \sigma_{c_j}) = \prod_i \mathcal{N}\left(f_\theta(p'_i) - h^j(p^j_i)|0, \sigma_{c_j}\right). \tag{8}$$

For N view registration, the final likelihood is computed as the product from N sets of contours:

$$l(\theta|C_1, ..., C_N, \sigma_{c_1}, ..., \sigma_{c_N}) = \prod_j^N l_j(\theta|C_j, \sigma_{c_j}). \tag{9}$$

2.4   Quantifying Uncertainty by Means of Sampling

In addition to measuring conventional mesh surface registration errors between the most likely reconstruction and the ground truth using both the average and the Hausdorff distance, the Bayesian sampling approach provides a means of measuring the uncertainty of the fit. We perform these measurements by considering all the samples that define our target distribution. We compute the variance of each vertex point across these samples using the Frobenius norm, where we first compute the mean and the covariance for each direction in $3D$ coordinate space. Instead of having one most probable reconstruction, we infer the reduction in uncertainty in different regions of the bone and visualise it as a colour map.

To visualise our sample distribution and to investigate how the contour lines influences the reconstruction, we take slices of the samples across the length of the bone, the $z-$plane.

3   Experimental Setup and Results

A SSM of the femur is trained as is described in Sect. 2.1 using 113 bone surfaces. We use the process in Sect. 2.2 to develop synthetic contour data and thereafter show the applicability of the method on femur images. Our image data set consists of ten images simulated with a point rendering method through the perspective projection of ten $3D$ meshes. Half of the images are generated from meshes that were part of the model's training data, the other half is out of the model span.

To perform our experiment we use the Scalismo framework (Basel 2015) which is an open-source library for statistical shape modelling and model-based image analysis in the programming language Scala. This software already covers broad use cases for face image interpretation and 3D surface modelling. The framework is still under extension to allow X-ray simulations.

The $3D$ vertices of the model mean surface are passed through the contour projector described in Sect. 2.2 and projected to $2D$ for each X-ray viewing angle (see Fig. 1).

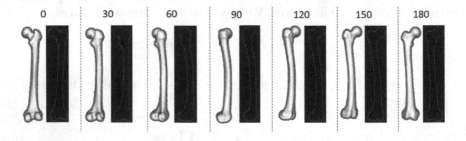

Fig. 1. Generation and projection of contour points from the 3D meshes.

The view-dependent contours are then fed as observations into the sampling algorithm described above. For a random proposal method, we consider 100 samples. From the resultant sample distribution (or surface meshes) we select one best posterior sample for each provided imaging angle. For the guided proposal method we use the best posterior of the random walk as the initial state of the model. We sample the step-size ϵ of the leapfrog step uniformly from

[0.000001, 0.0005). We also use a varying contour likelihood standard deviation σ to better represent different regions of the image. As such, areas with greater variance in the model have a higher σ. The number of steps denoted by L follows the Poisson distribution with Poisson(50). We use $n = 700$ HMC samples. The full experimental pipeline can be seen in the Fig. 2 below.

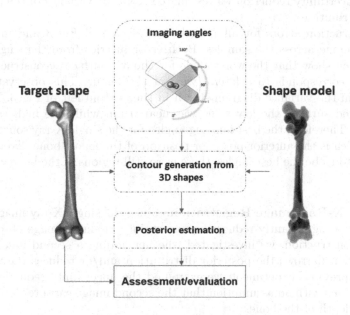

Fig. 2. Experimental setup.

We performed 2 experiments to evaluate different aspects of the proposed method. The first one evaluated the efficiency and accuracy of the proposed sampling algorithm with a single X-ray and the second one examined the feasibility of applying the method to a medical problem, that is $2D/3D$ reconstruction of bone surfaces with limited number of X-ray views [max n = 2].

3.1 2D/3D Reconstruction with Synthetic Data

Provided with an X-ray image contour projection, our task was to reconstruct a personalised $3D$ bone surface. Since the aim is to reduce the uncertainty in reconstruction using the limited number of X-ray images, it was important to assess the impact of the choice of the imaging angle in a setup with a single X-ray image and a pair of X-ray images (orthogonal to each other) on the quality and accuracy of the reconstructed surfaces.

Single X-Ray Registration. The single image experiment was performed to determine the optimal viewing direction for imaging a patient using a single X-ray source. The experiment was done for 7 different viewing directions with an angle α ranging from 0 and 180° in an increment of 30. For each α there was one corresponding X-ray image to which we then registered a model and obtained the posterior distribution. The most probable shape from the distribution was used to represent the inferred $3D$ bone surface. The error between the inferred $3D$ shape and the ground truth was estimated with surface distance measures. The localised uncertainty reduction was estimated from the variance of the posterior sample distribution.

Reconstruction errors for all 7 angles are shown in Fig. 3 and in Fig. 4 - with an average across the samples. Both error metrics from the single X-ray image setting show that the worst angle for the $3D$ surface reconstruction was 90°, which corresponds to a lateral view of the femur. This observation was confirmed in the variance diagram shown in the bottom left of Fig. 3. For each reconstructed surface, the low value was near 0 mm while the high value was near 5 mm. Therefore, the best imaging angles for the single X-ray source were 0 or 180° which is the anterior-posterior position of the femur bone. Even though these seemed to be the best angles, there were still regions of the bone with high variance.

Bi-Planar X-Ray Image Registration. Since the single X-ray image experiment did not significantly reduce the uncertainty in single image observations for our reconstruction, we investigated whether adding a second image to the reconstruction narrows the posterior distribution and/or reduces the variance. As for the previous experiment, we examined the uncertainty reduction in $3D$ reconstructions with an assumption that the second image captures information about the depth of the bone.

For this setup, an angle ϵ of separation between the images of the same pair was fixed to 90°. The angle α between the pairs of images ranged from 0 to 180° in increments of 30 in order to find the combination that leads to the optimal reconstruction, that is, a reconstruction closest to the ground truth and having the least variance. The top right of Fig. 3 shows that the average distance error decreased with the addition of a second image for all the imaging angles. This is consistent with our expectation as the second image captures the depth of the bone and reduces variance. This was also evident in the reconstructed femurs on the bottom right of Fig. 3 where the shaft has less variance. However, the femoral head and the condyles were still regions of high uncertainty, hence the Hausdorff distance remained high. For these regions of the bone contour extraction becomes more complex because of self occlusion at the ends of the femur. The greater trochanter is superimposed with some part of the head and the intertrochanter region. When the contour extraction is applied it is only the outer contours that are detected rather than the contours from the regions of great interest. In the $3D$ view these contour lines are not connected or continuous.

4   Conclusion

We have presented a novel approach to $3D$ shape reconstruction from a pair of X-ray image contour projections, with application to the femur bone. While tra-

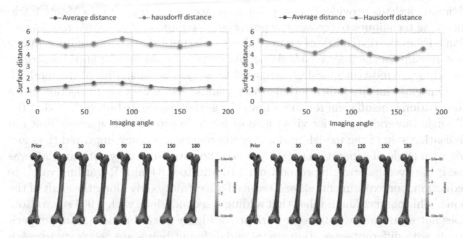

Fig. 3. Reconstruction results: the top row shows the surface distance between the reconstructed bone and the ground truth for a single X-ray experiment (left) and a bi-planar X-ray experiment (right), bottom row shows 3D reconstructed femurs for each viewing direction from a single X-ray experiment (left) and a bi-planar X-ray experiment (right). The colours represent the variance, red for high values and blue for low values. (Color figure online)

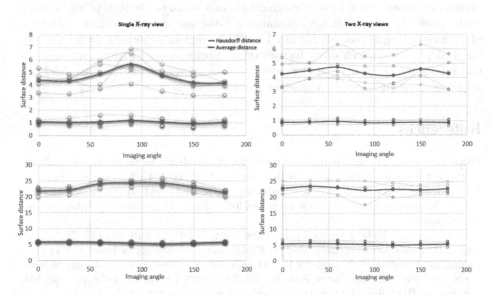

Fig. 4. Reconstruction results for a dataset with 10 femur bones: the top row shows the surface distance error between the reconstructed bone and the ground truth for the target images that are within the reach of the model, bottom row shows the surface distance error between the reconstructed bone and the ground truth for the target images that are out of the model reach. The bold colours shows an average across the samples while the light lines show each sample. (Color figure online)

ditional methods provide a single point-estimate, our approach uses Monte Carlo sampling techniques to recover the posterior distribution of plausible reconstructions. From the distribution, we were able to compute the amount of uncertainty that remains in the inferred $3D$ shape. Synthetic contour images and their corresponding ground-truth meshes were used to assess the feasibility of the method.

Our experiments indicate that using a single X-ray contour projection as an observation is insufficient for an accurate $3D$ surface estimate but, rather, that the $0°$ angle (anterior-posterior view) is most likely to provide a favourable first initialisation step of the model. Surface reconstruction accuracy improved the most when an angle-of-separation of $90°$ between the two X-ray views of 0 and $90°$ was used; the average error improved from $1.07\,\text{mm}$ to $0.84\,\text{mm}$. Regarding variance reduction, our experiments show a reduced uncertainty only along the shaft of the bone. This observation implies that adding a second view, with a $90°$ separation, does not significantly reduce the variance of highly curved regions. More experiments with different separation angles and different bones are necessary to reach a stronger conclusion regarding the suitability of our proposed method to 3D-2D reconstruction. Furthermore, since MC-based sampling allows different sources of information to be accommodated in its processing chain, our method can readily extend to intensity-based inferencing. We thus plan to investigate whether inner contours may help further constrain the model.

Once better established, we believe our proposed approach may benefit clinicians in two novel ways. First, the clinician is able to know the level of uncertainty with which the algorithm has estimated a bone surface. In addition, this type of understanding is possible for any local bone region; especially useful for pre-surgical planning which often focuses on a specific area. Finally, such insights can inform which viewing angles provide the least ambiguity, and are likely to improve the accuracy of bone surface reconstructions.

References

Baka, N., et al.: 2D–3D shape reconstruction of the distal femur from stereo X-ray imaging using statistical shape models. Med. Image Anal. **15**(6), 840850 (2011)

University of Basel. Scalismo—a scalable image analysis and shape modelling software framework available as open source (2015). https://github.com/unibas-gravis/scalismo

Duane, S., et al.: Hybrid Monte Carlo. Phys. Lett. B **195**(2), 216–222 (1987)

Egger, B.: Semantic Morphable Models. Ph.D. thesis. University of Basel (2017)

Fleute, M., Lavallée, S.: Nonrigid 3-D/2-D registration of images using statistical models. In: Taylor, C., Colchester, A. (eds.) MICCAI 1999. LNCS, vol. 1679, pp. 138–147. Springer, Heidelberg (1999). https://doi.org/10.1007/10704282_15

Hotelling, H.: Analysis of a complex of statistical variables into principal components. J. Educ. Psychol. **24**(6), 417 (1933)

Hurvitz, A., Joskowicz, L.: Registration of a CT-like atlas to uoroscopic X-ray images using intensity correspondences. Int. J. Comput. Assist. Radiol. Surg. **3**(6), 493 (2008)

Lüthi, M., et al.: Gaussian process morphable models. IEEE Trans. Pattern Anal. Mach. Intell. (2017)

Neal, R.M., et al.: MCMC using Hamiltonian dynamics. Handbook of Markov Chain Monte Carlo **2**(11), 2 (2011)

Sadowsky, O., Chintalapani, G., Taylor, R.H.: Deformable 2D-3D registration of the Pelvis with a limited field of view, using shape statistics. In: Ayache, N., Ourselin, S., Maeder, A. (eds.) MICCAI 2007. LNCS, vol. 4792, pp. 519–526. Springer, Heidelberg (2007). https://doi.org/10.1007/978-3-540-75759-7_63

Schönborn, S., et al.: Markov chain Monte Carlo for automated face image analysis. Int. J. Comput. Vision **123**(2), 160–183 (2017)

Suter, T., et al.: Viewing perspective malrotation inuences angular measurements on lateral radiographs of the scapula. J. Shoulder Elbow Surg. **29**(5), 1030–1039 (2020)

Tu, Z., et al.: Image parsing: unifying segmentation, detection, and recognition. Int. J. Comput. Vision **63**(2), 113–140 (2005)

Zhu, S.C., Zhang, R., Tu, Z.: Integrating bottom-up/top-down for object recognition by data driven Markov chain Monte Carlo. In: Proceedings of the IEEE Conference on Computer Vision and Pattern Recognition, vol. 1, pp. 738–745. IEEE (2000)

Zoppo, G.: Hamiltonian Monte Carlo for fitting point distribution models. Ph.D. thesis. University of Torino (2018)

Learning Shape Priors from Pieces

Dennis Madsen[✉], Jonathan Aellen, Andreas Morel-Forster, Thomas Vetter,
and Marcel Lüthi

Department of Mathematics and Computer Science, University of Basel, Basel,
Switzerland
{dennis.madsen,jonathan.aellen,andreas.forster,
thomas.vetter,marcel.luethi}@unibas.ch
https://gravis.dmi.unibas.ch

Abstract. Point Distribution Models (PDM) require a dataset in which
point-to-point correspondence between the individual shapes has been
established. However, in the medical domain, minimising radiation exposure and pathological deformations are reasons why healthy anatomies
are often only available as partial observations. To exploit the partial
shapes for learning shape models, previous methods required at least a
few complete shapes and, either a robust registration method or a robust
learning algorithm. Our proposed method implements the idea of multiple imputations from Bayesian statistics. We learn a PDM from a dataset
consisting of *only* incomplete shapes and a single full template. For this,
we first estimate the posterior distribution of point-to-point registrations
for each partial observation. Then we construct the PDM from the set
of registration distributions. We quantitatively evaluate our method on
a 2D dataset of hands and a 3D dataset of femurs with known ground-
truth. Furthermore, we showcase how to use our method on only partial
clinical data to build a 3D statistical model of the human skull. The code
is made open-source and the synthetic dataset publicly available.

Keywords: Statistical Shape Models · Point Distribution Models ·
Probabilistic registration · Multiple imputation

1 Introduction

Statistical Shape Models (SSMs) are a well-established tool for medical image
analysis. They can be used to automatically quantify whether a given shape is
anatomically normal, or how abnormal the shape is. This can e.g.. be used in
image segmentation tasks as regularisation. Also, SSMs can be used to reconstruct the complete shape from a partial observation, which is useful for forensic
investigation, reconstructive surgeries or patient-specific implant design.

Recently SSMs have been incorporated into deep learning pipelines as deformation regularisers. In [25], an SSM of the right ventricle (RV) chamber is used

---

Code available at https://github.com/unibas-gravis/shape-priors-from-pieces.

© Springer Nature Switzerland AG 2020
M. Reuter et al. (Eds.): ShapeMI 2020, LNCS 12474, pp. 30–43, 2020.
https://doi.org/10.1007/978-3-030-61056-2_3

in the loss function when training a U-NET. In [27], a fully connected neural network is trained to regress the parameters of an SSM from two cardiovascular magnetic resonance (CMR) image views and patient metadata. In [28], an SSM is used to learn a 2D to 3D mapping of liver images. A direct usage of an SSM is shown in [24], where SSMs are used to reconstruct the facial surface from 2D RGB images and the skull structure from CT images. The classical SSM can also be extended to incorporate patient-specific information [2]. As an alternative to performing a Principal Component Analysis (PCA) decomposition, SSMs can also be build to incorporate the non-linear relationship between shapes [1], while still maintaining performance capabilities as a linear model.

In this work, we make use of Point Distribution Models (PDM) which is a type of SSM. PDMs provide inherent correspondence, which we consider to be an important property for a lot of automatic analysis beyond segmentation. Other models without a point correspondence assumption, are e.g.. SSM based on level sets [23] or non-parametric shape priors [11].

In all of the above-mentioned papers, the PDMs are built from complete and healthy shapes in point-to-point correspondence. However, in the medical domain, either the data is captured because there is a pathology, or it is scanned only partially to capture the essential part of the structure while minimising the radiation danger. Hence, usually, only a part of the healthy anatomy is observed. Our paper is motivated by a practical example where we need to build a skull PDM from children Computed Tomography (CT) images. To minimise the radiation exposure during the image acquisition, only partial scans are taken depending on the area of interest. Several methods address the construction of PDMs where some of the training shapes are partial. In [10], each landmark is assigned a probability of being an outlier. This is used to compute a mean of the dataset where landmarks with lower probability of being an outlier have more influence on the mean shape. In [13], the training surfaces are divided into patches and each patch is assigned a probability of being an outlier. The outlier detection is performed with PCOut [3] which identifies samples that do not fit well into the distribution. Probabilistic PCA [22] is then used to iteratively build the PDM and replace the outlier parts with healthy parts. In [7], the shape model is computed using robust PCA (RPCA) to be able to marginally improve the model through partial data. In [15], they extend this idea to have a probability of being an outlier assigned for each landmark in a shape. The authors then extend their method to a robust kernel SSM, to have a non-linear model for better compactness [16]. All of the aforementioned methods use off-the-shelf registration methods and focus on building the models robustly from the noisy registration results. Often the registrations shrink substantially where parts are missing, to which the learning algorithm then has to be robust. The majority of the methods decompose the data matrix into a low-rank matrix containing correct data and a sparse matrix with the corrupted data. These matrices are mainly found via convex optimisation and usually require non-corrupted data to be present in the dataset.

In contrast, we introduce a principled way to go from *only* partial data observations to a PDM, without the need for multiple intermediate steps. We do not need to detect if parts are outliers or to assign weights for each landmark in a shape. Our method has its foundation in how missing data is handled in Bayesian statistics. There are three ways to handle missing data [8]:

- Discard observations with any missing values.
- Rely on the learning algorithm to handle missing values in the training phase.
- Impute all missing values (complete the data-matrix before using any learning algorithm).

From [4] we know that if we choose a reasonable missing-data model, the imputed dataset is likely to provide a more accurate estimate than a strategy which discards the data with missing values. In our setting where we are only working with partial data, it is not even a possibility to discard the data. For data imputation, a range of different strategies exist. The simplest strategy is to impute the missing values with the mean or the median of the non-missing values for that feature. This approach assumes that the missing values were known in the complete-data and will bias the variance of the dataset towards zero. One would, in the worst case, end up only learning the mean shape of the dataset and nothing about the shape variance. A better option for imputation is to perform regression based on the remaining dataset. This usually overestimates correlations which are then reflected by the model, as no uncertainty is given to the missing part. Instead of single imputation, we can make use of multiple imputations. Multiple imputation was initially introduced in [20] to fill out non-responses in surveys. If the data is arbitrary missing, the Markov chain Monte Carlo (MCMC) method can be used to create multiple imputations by simulating draws from a Bayesian predictive distribution given the partial data [26].

In this paper, we propose to combine probabilistic registration and the idea of multiple imputations to build PDMs purely from partial data. For each partially observed shape, we estimate the posterior distribution of registrations using an MCMC framework. The posterior distribution, which we obtain as a set of samples, reflects not only the uncertainty in the registration of the partial data, but at the same time also contains multiple completions. These completions can be seen as multiple imputations. The PDM can then be computed using standard PCA on the complete data-matrix which contains the imputed samples for each partial data item. Quantitative experiments are performed on a 2D dataset of hands and a 3D dataset of femurs with known ground-truth. Finally, we show how a 3D skull model can be build from partial data. The main contributions of this paper are:

- To the best of our knowledge, we present the first method that learns PDMs from *only* partial data.
- We show how to extend the classical Bayesian statistical method on missing data to point-to-point registration of partial data.
- We show that multiple data imputation creates PDMs with better specificity and generalisation than if single imputation is used.

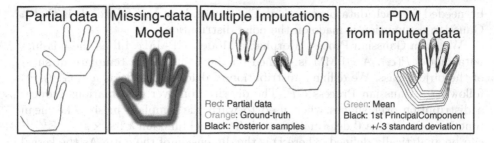

Fig. 1. Overview of our method. Given the partial data (red), the missing-data model (GPMM) is used to draw samples from the posterior distribution over registrations to impute the data. Finally, multiple imputations from each partial data item are used to compute the PDM. (Color figure online)

2   Method

A PDM is computed by performing PCA on the data matrix. As our dataset consists of partial observations, some of the entries are missing in the data matrix. Furthermore, we do not assume correspondence of the observed parts, so we do not even know the position in the matrix for the partial data. Our method implements the idea of multiple imputation from Bayesian statistics to complete the data matrix, while at the same time establishing correspondence in the dataset. For this, we independently process the M data items, each representing a partially observed shape. In contrast to previous methods, we are not only looking for the most likely imputation given a missing-data model but instead, we infer the full posterior distribution of completions

$$P(\boldsymbol{\alpha}|\boldsymbol{s}_p) = \frac{P(\boldsymbol{s}_p|\boldsymbol{\alpha})P(\boldsymbol{\alpha})}{\int P(\boldsymbol{s}_p|\boldsymbol{\alpha})P(\boldsymbol{\alpha})d\boldsymbol{\alpha}}, \tag{1}$$

where $\boldsymbol{s}_p$ denotes the partially observed shape and $\boldsymbol{\alpha}$ is the parameter vector controlling the imputation model. Using the full posterior, we account for the uncertainty of the registration *and* the reconstruction of missing areas when computing the PDM.

From the posterior distribution $P(\boldsymbol{\alpha}|\boldsymbol{s}_p)$, we randomly sample L imputations, such that the inferred data-matrix will be $X \in \mathbb{R}^{DN \times ML}$. Here N is the number of points which is multiplied with the dimensionality D of the embedding Euclidean space, $D = 2$ for the hand dataset and $D = 3$ for the femur and the skull datasets. By including multiple imputations into the PDM, we are able to take the uncertainty of the imputation under our missing-data model into account. An overview of our method is visualised in Fig. 1.

Alternatively, PCA can be performed directly on a set of distributions [6]. Instead of sampling L items from each distribution, they directly use the posterior distribution. This would help to scale the method if we are working in an even higher-dimensional space, where several thousands of imputations would be needed for each data observation. The downside of this approach is that a Gaussian assumption is made on the noise distribution.

We use a Gaussian Process Morphable Model (GPMM) [14] as the missing-data model $P(\alpha)$. A GPMM is defined on the domain of a template mesh s_t of the object class. We define a distribution of deformation fields $u \sim GP(\mu, k)$ following a Gaussian Process GP. The distribution over deformations induces a distribution over shapes when used to warp the template mesh. The mean function μ is set to a $\mathbf{0}$ deformation, and the kernel function $k : s_t \times s_t \to \mathbb{R}^{D \times D}$ can be analytically defined, where D is the dimension of the data. As the kernel function can be analytically defined, we do not even need a large dataset to estimate our missing-data model from. A simple covariance function to use (for 3D) is:

$$k(x, x') = \begin{bmatrix} g(x, x') & 0 & 0 \\ 0 & g(x, x') & 0 \\ 0 & 0 & g(x, x') \end{bmatrix} \tag{2}$$

with $g(x, x')$ being a Gaussian kernel

$$g(x, x') = s \cdot \exp(\frac{-\|x - x'\|^2}{\sigma^2}). \tag{3}$$

In the GP framework, multiple simple kernels can be combined to provide richer priors. A statistical covariance kernel estimated from data can be used or they can be based on expert knowledge about the shape deformations of a targeted class, such as smoothness or symmetry.

Furthermore, the template mesh can either be a single full data-item from the dataset, as in our hand and femur example, or it can be handcrafted, as in our skull model example. By choosing a handcrafted template and an expert-designed kernel, we can remove the need for even a single complete example in our skull experiments.

We use a low-rank approximation of the GP, based on the truncated Karhunen-Love expansion as in [14], in order to reduce the computational complexity of the model. This gives us a linear, parametric model ruled only by a coefficient vector $\alpha \sim \mathcal{N}(0, 1)$. The deformation model is then defined as

$$u[\alpha](x) = \mu(x) + \sum_{i=1}^{r} \alpha_i \sqrt{\lambda_i} \phi_i(x), \alpha_i \sim \mathcal{N}(0, 1), \tag{4}$$

where r is the number of retained basis functions used in the approximation and λ_i, ϕ_i are the i-th eigenvalue and eigenfunction of the covariance operator associated with the kernel function k.

The posterior distribution as described in Eq. (1) cannot be obtained analytically. But, we can evaluate the unnormalized posterior value for any shape described by the model coefficients α:

$$P(\alpha|s_p) \propto P(s_p|\alpha)P(\alpha). \tag{5}$$

Fig. 2. The dataset of hands with marked fingertips.

Using the Metropolis-Hastings (MH) algorithm [9], we can draw samples from the posterior, based only on the point-wise evaluation of the posterior.

For the MH algorithm, we only need to specify a likelihood function and a proposal generator. The prior $P(\alpha)$ is a standard multivariate normal distribution induced by the low-rank approximation. As likelihood model $P(s_p|\alpha)$, we use an independent point evaluator, as also used in [19]:

$$P(s_p|\alpha) = \prod_{i=1}^{n} \mathcal{N}(d_{l^2}(s_p^i, \alpha^i); 0, \sigma_{l^2}^2). \tag{6}$$

Here, d_{l^2} is the L2 distance between the i-th point from the target s_p^i and its closest point on the model instance denoted by α^i. We expect the observed deviations to follow a normal distribution with 0 mean and $\sigma_{l^2}^2$ variance.

The most common proposal distribution is the standard random-walk sampler. However, this is known to have slow convergence and long lag times (time between independent samples) in high dimensional spaces, as is our model space. To overcome this, we take advantage of the recent development in geometry-aware proposal strategies [17,18], which integrates correspondence estimation into the proposal step. For the exact details of the proposal distribution we refer to the provided implementation. Furthermore, we also make use of manually defined correspondences (clicked landmarks) to stabilise convergence.

For the partial data in this work, we assume that the data is a subset of the model, i.e. no overgrown regions from e.g.. metal implants or cancerous growth. Those artefacts would need to be manually removed and instead be completed by our methods.

3 Experiments

In the following experiments, we first demonstrate our method on a synthetic dataset of 2D hands and a 3D clinical dataset of femurs with artificially removed parts. As we have complete hands and femurs, we will compare our method with the ground-truth PDMs. PDMs (M) are usually evaluated based on three measurements: specificity S(M) (evaluate if the model only generates instances that are similar to those in the training set), generalisation G(M) (the ability to describe instances outside of the training set) and compactness C(M) (a model's ability to use a minimal set of parameters to represent the data variability) as described in [21]. From these measures, specificity and generalisation are the most important measures when using the model as prior information in other

(a) Mean, +/-3 (red/blue) standard deviation of the 4 first GPMM components.

(b) Random samples from the GPMM.

Fig. 3. Visualisation of the analytically defined 2D hand-GPMM. (Color figure online)

learning algorithms. We want the model to stay within the shape space, but also to be able to explain new data from the same shape space. For most of the experiments we do not plot the compactness as this is of less importance and very similar for all of the models. Finally, we apply our method to build a skull PDM from purely partial data.

The target meshes in all the experiments have been initially landmark aligned to a template mesh. For the synthetic experiments where parts are cut from the target meshes, we still keep the global alignment to avoid factoring pose difference into the model comparison.

All experiments are implemented in the open-source library Scalismo[1] and made publicly available. The only exception to the publicly available data and code is the partial skull dataset.

3.1   2D Synthetic Hand Experiment

For the 2D hand experiment, we make use of 12 synthetic 2D hand meshes as the targets, visualised in Fig. 2. An additional hand mesh is used as the reference mesh to construct a hand-GPMM as shown in Fig. 3. The computation of the generalisation measure for the hand models are computed by a "leave-one-out" approach. For the specificity measure, we use 1000 random samples.

The kernel used for creating the hand-GPMM is a mixture of smooth Gaussian kernels and an expert-designed kernel which is used to separately move the fingers from side to side. With the *finger* kernel we showcase how the incorporation of domain information can outperform standard kernels.

A Gaussian kernel favours smooth deformations with strong correlations between nearby points. However, this introduces a strong correlation between nearly touching sides of two neighbouring fingers, while at the same time both sides of one finger move more independently. The finger kernel aims at allowing side movements of a whole finger while preserving the overall finger shapes. An illustration of the finger kernel is visualised in Fig. 4.

We define the finger kernel for each finger individually. For this, we mark the start of each finger on both sides as A and B as well as the fingertip as C. We then define the kernel as:

$$k(x, x') = s(x, x') \Sigma e^{-\frac{1}{2\sigma^2} d_{ABC}(x,x')^2} . \tag{7}$$

---

[1] https://scalismo.org.

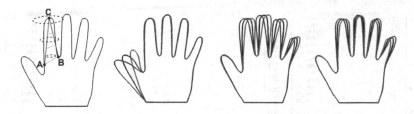

Fig. 4. Illustration of the expert-designed finger kernel. The left image shows the helper points to construct this kernel for the index finger: helper lines (red), lines connecting x and x' where $d_{ABC}(x, x') = 0$ (orange) and illustrations of the scaled covariance matrices $s(x, x')\Sigma_i$ (blue) as overlays. The three visualisations on the right show how the first three principal components of a finger kernel only model encode side-ways moving fingers. (Color figure online)

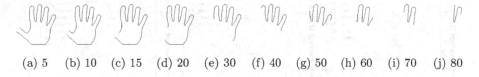

(a) 5   (b) 10   (c) 15   (d) 20   (e) 30   (f) 40   (g) 50   (h) 60   (i) 70   (j) 80

Fig. 5. Hand example of partial data. The caption lists how much of each shape has been removed (5–80%) starting from the fingertip of the thumb.

Given the closest points of x and x' on the set of lines AC and BC, $d_{ABC}(x, x')$ denotes the difference of the first entry in their barycentric coordinates. The covariance Σ is constructed by first defining a 2×2 diagonal matrix, in our example using the values 0.1γ and γ respectively. We then rotate the diagonal matrix by the angle of the x-axis and the direction MC, with M as the mid-point of AB. The scaling function $s(x, x')$ is the barycentric interpolated value between 0 and 1 using the product of the barycentric coordinates of the closest points, or 0 if x and x' do not map to the same finger. The complete expert-designed kernel is then the sum over all five finger kernels specified on the template.

Our choices of γ and σ can be seen in the published code.

Missing Finger Experiments. We clip the hands' dataset starting from a landmark on the top of each finger and cut away increasing amount (5–90%), as visualised in Fig. 5. We perform 2 different experiments with this setup. In the first experiment, a random finger is increasingly cut away from each of the hands. In the second experiment, all 12 hands are increasingly missing the thumb. In Fig. 6 we show the model measures of the missing thumb experiment. Due to space constraints, we only show the measures from this experiment, as the results for both of these experiments are very similar. We see that the models computed from multiple imputations are able to generalise much better than when using the Maximum a posteriori estimations (MAP) or mean solutions. Note, that the curve flattens for the MAP and the mean experiments as only 12

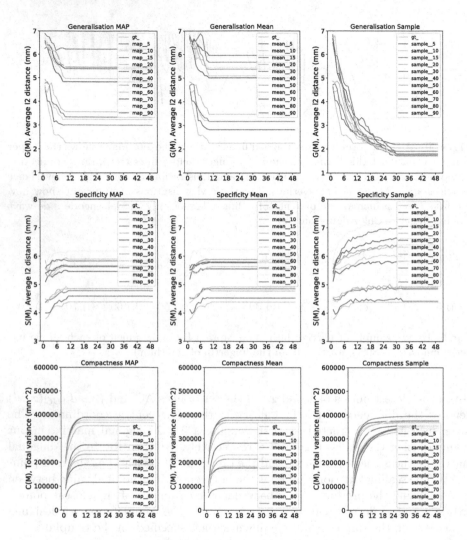

Fig. 6. Hand model measures. The left column shows the results for models computed with the Maximum a posteriori estimation (MAP) from each posterior distribution. The middle column is computed using the mean mesh from each posterior distribution. The right column shows the results using 100 imputations from each of the 12 posterior distribution. The last-mentioned model is clearly superior.

meshes are used to compute these models, whereas we use 100 imputations for each of the 12 target meshes in the multiple imputation experiment. If less than 30% of the meshes are missing, then we also obtain a better specificity measure compared to the MAP and mean solutions. Finally, we see that the multiple imputations are able to maintain much more of the total variance within the model when large parts of the targets are missing.

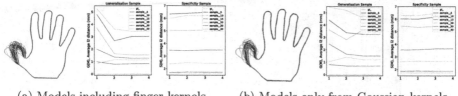

(a) Models including finger kernels.   (b) Models only from Gaussian kernels.

Fig. 7. Finger kernel ablation study. Red: Partial shape, Orange: Ground-truth shape, Green: MAP, Blue: Mean, Black: Random samples. The model using the finger kernel performs better than the purely Gaussian kernel model. (Color figure online)

Finger Kernel Ablation Study. In the ablation study, we demonstrate that the inclusion of expert knowledge into the kernel design improves performance slightly. We use a dataset of 5 identical meshes where only a small scaling difference is applied. When computing the generalisation measure from a leave-one-out approach, we therefore, know that the ground-truth model will be able to perfectly describe the mesh which is not in the model. From Fig. 7 we see that even though the posterior samples from the models look similar, the model with the finger kernels overall generalises better. Especially when a large part of the target is missing (>15%). The specificity measures are also marginally better by using finger kernels. The main takeaway from the ablation study is (not surprisingly), that the more expert knowledge we put into the design of a missing-data model, the better imputations we get.

3.2 3D Femur Experiment with Ground-Truth

For the initial femur-GPMM, we use a smooth Gaussian kernel as in [17]. In this experiment, we assume no additional domain knowledge. The template mesh and random samples from the femur-GPMM are shown in Fig. 8a. We use the publicly available dataset of 50 complete femur meshes that were extracted from computed tomography images[2]. We choose ten complete femur meshes as our training data, from where we clip a varying amount (5–25%) at different landmark locations. For eight of the bones, this clipping was done in a single location. For the remaining two, we spread the removal over four locations. The partial dataset is visualised in Fig. 8b. Similar to before we build PDMs from multiple imputations, the MAP solution and the mean and compare them additionally to the ground-truth PDM. For the specificity computation, the ground-truth registrations are used (these are also used to build the ground-truth PDM). For the generalisation, we use 40 registered meshes which are not included the dataset of the PDMs.

In Fig. 10 we see an example of a target shape (red) with multiple different imputations. We see how both the mean and the MAP solutions fail to perfectly predict the true shape. From the random samples we see that there is a broad

---

[2] Available at the SICAS Medical Image Repository [12].

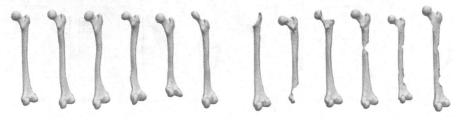

(a) Template (left) and random samples.   (b) Examples of the partial femurs.

Fig. 8. Illustration of random femur-GPMM samples and the used partial data.

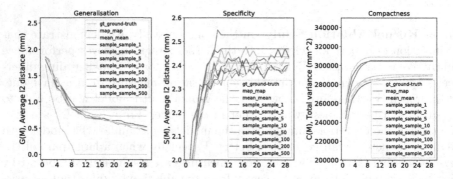

Fig. 9. Femur model measures. Multiple imputations lead to a better generalisation, while also keeping a better specificity than models built from the MAP or mean imputations.

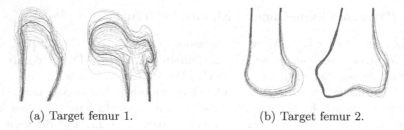

(a) Target femur 1.   (b) Target femur 2.

Fig. 10. Posterior distribution visualisations of the two first partial femurs from Fig. 8b. Red: Partial shape, Orange: Ground-truth shape, Green: MAP, Blue: Mean, Black: Posterior samples. It can be seen that the posterior samples better cover the ground-truth than both, the MAP and mean shape. (Color figure online)

distribution of possible imputations for the missing part. In Fig. 9 we compare PDMs created from the MAP, mean and multiple imputations. In the case of multiple imputations, we compare models created with a different number of samples. From the given example, we see that by using 5 or more imputations, we are able to create PDMs that generalise better, than by using the MAP or the mean solutions, while also maintaining better specificity.

3.3   3D Skull Model from Real Partial Data

In this experiment, we build a skull model from 16 partial skull pieces. For the skull-GPMM, we use a mixture of smooth Gaussian kernels and a symmetrical kernel around the sagittal plane [5]. As we have no ground-truth data, we can only qualitatively evaluate the model. In Fig. 11, examples of the partial data is shown as well as the hand-crafted skull template from which we create a skull-GPMM. We check the model deformations by varying the individual principal components. We see that the first principal component captures the size of the skull, which would also be expected.

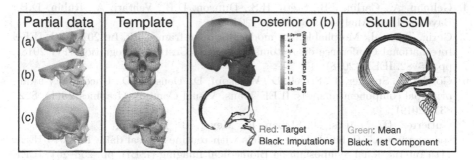

Fig. 11. Skull model experiment. The colour mesh is the point-wise sum of variances. As expected, a larger uncertainty (red) is inferred for the missing part. The visualisation of samples from the posterior in the 2D slice, show the remaining variability (black). To the right, the variation of the 1st principal component from −3 to 3 standard deviation of the resulting PDM is shown. (Color figure online)

4   Conclusion

In this paper, we introduced a principled method to create PDMs from *only* partial data observations. Our method implements the idea of multiple imputation from Bayesian statistics for point-to-point registration of partial data. With this, we can create multiple imputations of a single partial data observation to span the data-matrix. To the best of our knowledge, we are the first to create PDMs from multiple imputations of partial data. We show how this improves model specificity and generalisation. We can influence the imputations by choosing the missing-data model in the MCMC framework. Using Gaussian Process Morphable Models, we can not only design the model in an analytical way to encode smoothness or symmetry but also gradually include more and more data if complete shapes become available. Finally, we showed that our method is not only theoretically nice but that it can also be used in practice to create a PDM from only skull pieces.

Acknowledgements. This research is sponsored by the Gebert Rüf Foundation under the project GRS-029/17.

References

1. Ambellan, F., Zachow, S., von Tycowicz, C.: A surface-theoretic approach for statistical shape modeling. In: Shen, D., et al. (eds.) MICCAI 2019. LNCS, vol. 11767, pp. 21–29. Springer, Cham (2019). https://doi.org/10.1007/978-3-030-32251-9_3
2. Egger, B., Schirmer, M.D., Dubost, F., Nardin, M.J., Rost, N.S., Golland, P.: Patient-specific conditional joint models of shape, image features and clinical indicators. In: Shen, D., et al. (eds.) MICCAI 2019. LNCS, vol. 11767, pp. 93–101. Springer, Cham (2019). https://doi.org/10.1007/978-3-030-32251-9_11
3. Filzmoser, P., Maronna, R., Werner, M.: Outlier identification in high dimensions. Comput. Stat. Data Anal. **52**(3), 1694–1711 (2008)
4. Gelman, A., Carlin, J.B., Stern, H.S., Dunson, D.B., Vehtari, A., Rubin, D.B.: Bayesian Data Analysis. CRC Press (2013)
5. Gerig, T., et al.: Morphable face models-an open framework. In: 2018 13th IEEE International Conference on Automatic Face & Gesture Recognition (FG 2018), pp. 75–82. IEEE (2018)
6. Görtler, J., Spinner, T., Streeb, D., Weiskopf, D., Deussen, O.: Uncertainty-aware principal component analysis. IEEE Trans. Visual Comput. Graphics **26**(1), 822–831 (2019)
7. Gutierrez, B., Mateus, D., Shiban, E., Meyer, B., Lehmberg, J., Navab, N.: A sparse approach to build shape models with routine clinical data. In: 2014 IEEE 11th International Symposium on Biomedical Imaging (ISBI), pp. 258–261. IEEE (2014)
8. Hastie, T., Tibshirani, R., Friedman, J.: The Elements of Statistical Learning: Data Mining, Inference, and Prediction. Springer, New York (2009)
9. Hastings, W.K.: Monte Carlo sampling methods using Markov chains and their applications (1970)
10. Jingting, M., Lentzen, K., Honsdorf, J., Feng, L., Erdt, M.: Statistical shape modeling from Gaussian distributed incomplete data for image segmentation. In: Oyarzun Laura, C., et al. (eds.) CLIP 2015. LNCS, vol. 9401, pp. 113–121. Springer, Cham (2016). https://doi.org/10.1007/978-3-319-31808-0_14
11. Kim, J., Çetin, M., Willsky, A.S.: Nonparametric shape priors for active contour-based image segmentation. Sig. Process. **87**(12), 3021–3044 (2007)
12. Kistler, M., Bonaretti, S., Pfahrer, M., Niklaus, R., Büchler, P.: The virtual skeleton database: an open access repository for biomedical research and collaboration. J. Med. Internet Res. **15**(11), e245 (2013)
13. Lüthi, M., Albrecht, T., Vetter, T.: Building shape models from lousy data. In: Yang, G.-Z., Hawkes, D., Rueckert, D., Noble, A., Taylor, C. (eds.) MICCAI 2009. LNCS, vol. 5762, pp. 1–8. Springer, Heidelberg (2009). https://doi.org/10.1007/978-3-642-04271-3_1
14. Lüthi, M., Gerig, T., Jud, C., Vetter, T.: Gaussian process morphable models. IEEE Trans. Pattern Anal. Mach. Intell. **40**(8), 1860–1873 (2017)
15. Ma, J., Lin, F., Honsdorf, J., Lentzen, K., Wesarg, S., Erdt, M.: Weighted robust PCA for statistical shape modeling. In: Zheng, G., Liao, H., Jannin, P., Cattin, P., Lee, S.-L. (eds.) MIAR 2016. LNCS, vol. 9805, pp. 343–353. Springer, Cham (2016). https://doi.org/10.1007/978-3-319-43775-0_31
16. Ma, J., Wang, A., Lin, F., Wesarg, S., Erdt, M.: A novel robust kernel principal component analysis for nonlinear statistical shape modeling from erroneous data. Comput. Med. Imaging Graph. **77**, 101638 (2019)

17. Madsen, D., Morel-Forster, A., Kahr, P., Rahbani, D., Vetter, T., Lüthi, M.: A closest point proposal for MCMC-based probabilistic surface registration. In: Proceedings of the European Conference on Computer Vision (ECCV), August 2020

18. Madsen, D., Vetter, T., Lüthi, M.: Probabilistic Surface Reconstruction with Unknown Correspondence. In: Greenspan, H., et al. (eds.) CLIP/UNSURE -2019. LNCS, vol. 11840, pp. 3–11. Springer, Cham (2019). https://doi.org/10.1007/978-3-030-32689-0_1

19. Morel-Forster, A., Gerig, T., Lüthi, M., Vetter, T.: Probabilistic fitting of active shape models. In: Reuter, M., Wachinger, C., Lombaert, H., Paniagua, B., Lüthi, M., Egger, B. (eds.) ShapeMI 2018. LNCS, vol. 11167, pp. 137–146. Springer, Cham (2018). https://doi.org/10.1007/978-3-030-04747-4_13

20. Rubin, D.B.: Multiple imputations in sample surveys-a phenomenological Bayesian approach to nonresponse. In: Proceedings of the Survey Research Methods Section of the American Statistical Association, vol. 1, pp. 20–34. American Statistical Association (1978)

21. Styner, M.A., et al.: Evaluation of 3D correspondence methods for model building. In: Taylor, C., Noble, J.A. (eds.) IPMI 2003. LNCS, vol. 2732, pp. 63–75. Springer, Heidelberg (2003). https://doi.org/10.1007/978-3-540-45087-0_6

22. Tipping, M.E., Bishop, C.M.: Probabilistic principal component analysis. J. R. Stat. Soc. Ser. B (Stat. Methodol.) **61**(3), 611–622 (1999)

23. Tsai, A., et al.: A shape-based approach to the segmentation of medical imagery using level sets. IEEE Trans. Med. Imaging **22**(2), 137–154 (2003)

24. Xiao, D., et al.: Estimating reference bony shape model for personalized surgical reconstruction of posttraumatic facial defects. In: Shen, D., et al. (eds.) MICCAI 2019. LNCS, vol. 11768, pp. 327–335. Springer, Cham (2019). https://doi.org/10.1007/978-3-030-32254-0_37

25. Yang, H., Liu, Z., Yang, X.: Right ventricle segmentation in short-axis MRI using a shape constrained dense connected U-Net. In: Shen, D., et al. (eds.) MICCAI 2019. LNCS, vol. 11765, pp. 532–540. Springer, Cham (2019). https://doi.org/10.1007/978-3-030-32245-8_59

26. Yuan, Y.C.: Multiple imputation for missing data: concepts and new development. In: Proceedings of the Twenty-Fifth Annual SAS Users Group International Conference, vol. 267 (2000)

27. Yue, Q., Luo, X., Ye, Q., Xu, L., Zhuang, X.: Cardiac segmentation from LGE MRI using deep neural network incorporating shape and spatial priors. In: Shen, D., et al. (eds.) MICCAI 2019. LNCS, vol. 11765, pp. 559–567. Springer, Cham (2019). https://doi.org/10.1007/978-3-030-32245-8_62

28. Zhou, X.-Y., Wang, Z.-Y., Li, P., Zheng, J.-Q., Yang, G.-Z.: One-stage shape instantiation from a single 2D image to 3D point cloud. In: Shen, D., et al. (eds.) MICCAI 2019. LNCS, vol. 11767, pp. 30–38. Springer, Cham (2019). https://doi.org/10.1007/978-3-030-32251-9_4

Bi-invariant Two-Sample Tests in Lie Groups for Shape Analysis
Data from the Alzheimer's Disease Neuroimaging Initiative

Martin Hanik[(✉)][iD], Hans-Christian Hege[iD], and Christoph von Tycowicz[iD]

Zuse Institute Berlin, Berlin, Germany
{hanik,hege,vontycowicz}@zib.de

Abstract. We propose generalizations of the Hotelling's T^2 statistic and the Bhattacharayya distance for data taking values in Lie groups. A key feature of the derived measures is that they are compatible with the group structure even for manifolds that do not admit any bi-invariant metric. This property, e.g. assures analysis that does not depend on the reference shape, thus, preventing bias due to arbitrary choices thereof. Furthermore, the generalizations agree with the common definitions for the special case of flat vector spaces guaranteeing consistency. Employing a permutation test setup, we further obtain nonparametric, two-sample testing procedures that themselves are bi-invariant and consistent. We validate our method in group tests revealing significant differences in hippocampal shape between individuals with mild cognitive impairment and normal controls.

Keywords: Non-metric shape analysis · Lie groups · Geometric statistics

1 Introduction

Shape analysis is applied successfully in a variety of different fields and is further fuelled by the ongoing advance of 3D imaging technology [2]. Although the objects themselves are embedded in Euclidean space the resulting shape data is often part of a complex nonlinear manifold. Thus, methods for its analysis must generalize Euclidean statistical tools. Lie groups form the natural domain of shapes when they are modeled as transformations between different subjects and a common reference or atlas and the idea to represent them entirely via these transformations has been very successful since its introduction by D'Arcy Thompson over 100 years ago [22]. Indeed, there are various different models that consider different Lie groups. While configurations of the human spine can be encoded in the low dimensional groups of translations and rotations [1,6], the large deformation diffeomorphic metric mapping framework (LDDMM) [11,14] represents deformations of images in the infinite-dimensional group of diffeomorphisms. The classical matrix groups also appear in physics based shape

© Springer Nature Switzerland AG 2020
M. Reuter et al. (Eds.): ShapeMI 2020, LNCS 12474, pp. 44–54, 2020.
https://doi.org/10.1007/978-3-030-61056-2_4

spaces [4,23], diffusion tensor imaging [13] and in the characterization of volume [24] and surface [3] deformations.

Geometrically defined statistical methods in Riemannian manifolds have long been considered and they provide powerful tools not only for shape analysis [18]. For Lie groups, however, they do not respect the group structure as they are only invariant with respect to left and right translations, as well as inversion, when there exists a bi-invariant metric. Important examples, where this is not the case, are the group of rigid-body transformations and the general linear group for dimensions greater than one. To overcome these problems, Pennec and Arsigny generalized the notions of the mean, covariance and Mahalanobis distance in a bi-invariant way [19]. We build upon and extend their work to derive bi-invariant generalizations of the Hotelling's T^2 statistic and Bhattacharayya distance for observations taking values in Lie groups. These then induce two-sample permutation tests that are themselves compatible with the group structure even for manifolds that do not admit any bi-invariant metric. Our generalizations are consistent in that they agree with the original expressions in flat vector spaces; this is not true for previous generalizations in Riemannian manifolds [12,16]. We evaluate the proposed group test for the morphometric analysis of pathological malformations associated to cognitive decline, viz. mild cognitive impairment, which is common in the elderly and represents an intermediate stage between normal cognition and Alzheimer's disease.

2   Theoretical Background

Basics of Lie Groups. In the following, we give a short summary of the theory of Lie groups. For more information see for example [20]. Additional information on differential geometry can be found in [7]. In the following we use "smooth" synonymously with "infinitely often differentiable".

A Lie group G is a smooth manifold that has a compatible group structure, that is, there is an identity element $e \in G$ and a smooth, associative (not necessarily commutative) map $G \times G \ni (g, h) \mapsto gh \in G$ as well as a smooth inversion map $G \ni g \mapsto g^{-1}$. An example of a Lie group is the general linear group GL(n), i.e. the set of all bijective linear mappings on a vector space V, where the group operation is the composition of mappings (i.e. a matrix multiplication), with e being the identity map. Whenever we speak of matrix groups in the following, arbitrary subgroups of GL(n) are meant. For each $g \in G$ the group operation defines two automorphisms on G: the left and right translation $L_g : h \mapsto gh$ and $R_g : h \mapsto hg$. Their derivatives $d_h L_g$ and $d_h R_g$ at $h \in G$ map tangent vectors $X \in T_h G$ bijectively to the tangent spaces $T_{gh} G$ and $T_{hg} G$, respectively. In particular, it holds that $T_g G = \{d_e L_g(X) : X \in T_e G\} = \{d_e R_g(X) : X \in T_e G\}$. Thus, each X in $T_e G$ determines a vector field $\widetilde{X}$ by $\widetilde{X}_g = d_e L_g(X)$ for all $g \in G$. It is called left invariant because $\widetilde{X}_{L_g(h)} = d_h L_g(\widetilde{X}_h)$ for all $h \in G$, that is, the value at a left translated point is the left translated vector. Furthermore, the converse also holds: every left invariant vector field is uniquely determined by its value at the identity. For matrix groups with identity matrix I we get the simple equation $d_I L_A(M) = AM$ for an element A and a matrix M in the tangent space at I. Right invariant vector fields are defined analogously and have parallel properties.

The integral curve $\alpha_X : \mathbb{R} \to G$ of an invariant (left or right) vector field $\widetilde{X}$ with $X = \widetilde{X}_e$ determines a 1-parameter subgroup of G through e since $\alpha_X(s + t) = \alpha_X(s)\alpha_X(t)$ for all $s, t \in \mathbb{R}$. The *group exponential* exp is then defined by $\exp(X) = \alpha_X(1)$. It is a diffeomorphism in a neighbourhood of e and, hence, we can also define the *group logarithm* log as its inverse there. In the case of matrix groups they coincide with the matrix exponential and logarithm.

Given two vector fields X, Y on G a so-called connection ∇ yields a way to differentiate Y along X; the result is again a vector field which we denote by $\nabla_X Y$. With $\gamma' := \frac{d\gamma}{dt}$ we can then define a geodesic $\gamma : [0, 1] \to G$ by $\nabla_{\gamma'}\gamma' = 0$ as a curve without acceleration. An important fact is that every point $g \in G$ has a so-called normal convex neighbourhood U. Each pair $f, h \in U$ can be joined by a unique geodesic $[0, 1] \ni t \mapsto \gamma(t; f, h)$ that lies completely in U. Furthermore, with $\gamma'(0; g, h) = X$, this defines the exponential $\mathrm{Exp}_g : T_g G \to G$ at g by $\mathrm{Exp}_g(X) := \gamma(1; g, h)$. It is also a local diffeomorphism with local inverse $\mathrm{Log}_g(h) = \gamma'(0; g, h)$. If the so-called Levi-Civita connection is used, then Exp and Log are called *Riemannian exponential* and *logarithm*, respectively. The Riemannian and group maps coincide if and only if G admits a bi-invariant Riemannian metric, that is, a smoothly varying inner product on the tangent spaces that is invariant under left *and* right translations.

If we endow G with a Cartan-Shouten connection [8], then geodesics and left (or right) translated 1-parameter subgroups coincide. Thus, for every $g \in G$ there is also a normal convex neighbourhood U such that the map $U \ni h \mapsto \log_g(h) = d_e L_g \log(g^{-1}h)$ is well-defined. It can be interpreted as the "difference of h and g" taken in $T_g G$. For the rest of the paper we will assume that we work in such a neighborhood U.

Another important automorphism of G is the conjugation $C_g : h \mapsto ghg^{-1}$. Its differential w.r.t h is called the group adjoint and denoted by $\mathrm{Ad}(g)$. It acts on vectors $X \in T_e G$ by

$$\mathrm{Ad}(g)X = d_{g^{-1}}L_g(d_e R_{g^{-1}}(X)) = d_g R_{g^{-1}}(d_e L_g(X)).$$

For matrix groups this reduces to $\mathrm{Ad}(A)(M) = AMA^{-1}$ for elements A and matrices M in the tangent space at the identity.

Hotelling T^2 Statistic for Riemannian Manifolds. Hotelling's T^2 test is the multivariate counterpart to the t-test. Given two data sets $(p_1, \ldots, p_m)$ and $(q_1, \ldots, q_n)$ in $\mathbb{R}^d$ with means $\bar{p}$ and $\bar{q}$, the data's pooled sample covariance is given by

$$S = \frac{\sum_{i=1}^m (p_i - \bar{p})(p_i - \bar{p})^T + \sum_{j=1}^n (q_j - \bar{q})(q_j - \bar{q})^T}{m + n - 2}.$$

The Hotelling T^2 statistic is then defined as the square of the Mahalanobis distance scaled with $mn/(m + n)$:

$$t^2(\{p_i\}, \{q_i\}) = \frac{mn}{m + n}(\bar{p} - \bar{q})^T S^{-1}(\bar{p} - \bar{q}).$$

It measures the difference of $\bar{p}$ and $\bar{q}$ weighted against the inverse of the pooled covariance. Therefore, directions in which high variability was observed are weighted less than those with little spreading around the corresponding component of the mean.

In [16, Sect. 3.3] Muralidharan and Fletcher introduce a generalization of the T^2 statistic to Riemannian manifolds M, i.e. for samples $(p_1, \ldots, p_m)$, $(q_1, \ldots, q_n)$ in M. The centers of the data sets are then given by the Fréchet means $\bar{p}, \bar{q} \in M$, respectively. Assuming that $\bar{p}, \bar{q}$ are unique, the difference between the means can be replaced by the Riemannian logarithms $v_{\bar{p}} = \mathrm{Log}_{\bar{p}}(\bar{q}) \in T_{\bar{p}}M$ or $v_{\bar{q}} = \mathrm{Log}_{\bar{q}}(\bar{p}) \in T_{\bar{q}}M$. Depending on the choice, the vectors are from different tangent spaces. Analogously the covariance matrices can be defined by

$$W_{p_i} = \frac{1}{m} \sum_{i=1}^{m} \mathrm{Log}_{\bar{p}}(p_i)\mathrm{Log}_{\bar{p}}(p_i)^T,$$

$$W_{q_i} = \frac{1}{n} \sum_{i=1}^{n} \mathrm{Log}_{\bar{q}}(q_i)\mathrm{Log}_{\bar{q}}(q_i)^T.$$

Since there is no canonical way to compare vectors from different tangent spaces, Muralidharan and Fletcher propose to calculate a generalized T^2 statistic at both means and average the results. This leads to the generalized T^2 statistic

$$t^2(\{p_i\}, \{q_i\}) = \frac{1}{2}\left(v_{\bar{p}}^T W_{p_i}^{-1} v_{\bar{p}} + v_{\bar{q}}^T W_{q_i}^{-1} v_{\bar{q}}\right)$$

for Riemannian manifolds.

3   Group Testing in Lie Groups

3.1   Bi-invariant Mahalanobis Distance

In [19] Pennec and Arsigny define a *bi-invariant mean* on a Lie group G of dimension $k \in \mathbb{N}$ and then show that there is a canonical way to generalize the notion of Mahalanobis distance to the Lie group setting. Given data $(g_1, \ldots, g_m)$ in a normal convex neighborhood, the bi-invariant mean $\bar{g}$ is defined implicitly as the solution of the group barycentric equation

$$\sum_{i=1}^{m} \log(\bar{g}^{-1} g_i) = 0.$$

It is equivariant with respect to left and right translations as well as inversion, i.e., for all $f \in G$ the means of left translated data $(fg_1, \ldots, fg_m)$, right-translated data $(g_1 f, \ldots, g_m f)$ and inverted data $(g_1^{-1}, \ldots, g_m^{-1})$ are $f\bar{g}$, $\bar{g}f$ and $\bar{g}^{-1}$, respectively [19, Thm. 11]. Bi-invariant means can be computed efficiently with a fixed point iteration [19, Alg. 1]. Pennec and Arsigny define the intrinsic (i.e., independent of the choice of coordinates) covariance tensor of the data at $\bar{g}$ by

$$\Sigma_{g_i} = \frac{1}{n} \sum_{i=1}^{n} \log_{\bar{g}}(g_i) \otimes \log_{\bar{g}}(g_i) \in T_{\bar{g}}G \otimes T_{\bar{g}}G,$$

where the tensor product $\otimes$ means that in any basis of $T_{\bar{g}}G$, the entries are $[\Sigma_{g_i}]^{ij} = 1/m \sum_l [\log_{\bar{g}}(g_l)]^i [\log_{\bar{g}}(g_l)]^j$. From this, the bi-invariant Mahalanobis distance of $f \in G$ to the distribution of the g_i can be defined by

$$\mu^2_{(\bar{g},\Sigma_{g_i})}(f) := \sum_{i,j=1}^{k} [\log_{\bar{g}}(f)]^i [\Sigma_{g_i}^{-1}]_{ij} [\log_{\bar{g}}(f)]^j, \tag{1}$$

where $[\Sigma_{g_i}^{-1}]_{ij}$ denotes the elements of the inverse of Σ_{g_i} in a given basis. It is left and right invariant because both translations amount to a joint change of basis of $\log_{\bar{g}}(g_i)$ and Σ_{g_i} whose effect cancels out because of the inversion of the covariance matrix in (1); see [17, p. 181].

3.2   Generalized Hotelling's T^2 Test

In this section we use the bi-invariant Mahalanobis distance from the previous section to define a bi-invariant generalization of the Hotelling T^2 statistic for data in Lie groups G of dimension $k \in \mathbb{N}$. First, note that we can always jointly translate the data such that the new mean is the identity e without changing Mahalanobis distances. Thus, instead of (1) we use the equivalent form

$$\mu^2_{(\bar{g},\Sigma_{g_i})}(f) = \sum_{i,j=1}^{k} [\log(\bar{g}^{-1}f)]^i [\widetilde{\Sigma}_{g_i}^{-1}]_{ij} [\log(\bar{g}^{-1}f)]^j$$

in the following, where

$$[\widetilde{\Sigma}_{g_i}]^{ij} := \frac{1}{m} \sum_{l=1}^{m} [\log(\bar{g}^{-1}g_l)]^i [\log(\bar{g}^{-1}g_l)]^j$$

is the *centralized covariance* of $(g_1, \ldots, g_m)$. This motivates the definition of the pooled covariance at the identity.

Definition 1. *Given data sets $(g_1, \ldots, g_m)$ and $(h_1, \ldots, h_n)$ in a Lie group G with bi-invariant means $\bar{g}$ and $\bar{h}$, their* pooled covariance *is defined by*

$$\widehat{\Sigma} := \frac{1}{m+n-2} \left(m\widetilde{\Sigma}_{g_i} + n\widetilde{\Sigma}_{h_i} \right).$$

With this, we propose the following generalization of the T^2 statistic for Lie groups.

Definition 2. *Given data sets* $(g_1, \ldots, g_m)$ *and* $(h_1, \ldots, h_n)$ *in a Lie group* G *with bi-invariant means* $\overline{g}$ *and* $\overline{h}$, *the* bi-invariant Hotelling's T^2 statistic *is defined by*

$$t^2(\{g_i\}, \{h_i\}) := \frac{mn}{m+n} \mu^2_{(e, \widehat{\Sigma})} \left(\overline{g}^{-1} \overline{h} \right).$$

Note that we could replace left by right translations in all definitions in this section. The resulting centralized and pooled covariance will be different in general, but the bi-invariant T^2 statistic turns out to be the same as translation effects cancel out.

3.3 Bhattacharyya Distance

Another index suggested for assessing the dissimilarity between two distributions that is also related to the Mahalanobis distance is the Bhattacharyya distance [5]. Given two data sets $(p_1, \ldots, p_m)$ and $(q_1, \ldots, q_n)$ in $\mathbb{R}^d$ with means $\overline{p}, \overline{q}$ and sample covariance S_{p_i}, S_{q_i}, the distance is defined as

$$D_B((\overline{p}, S_{p_i}), (\overline{q}, S_{q_i})) := \frac{1}{8}(\overline{p} - \overline{q})^T S^{-1}(\overline{p} - \overline{q}) + \frac{1}{2}\ln\left(\frac{|S|}{\sqrt{|S_{p_i}||S_{q_i}|}}\right),$$

where $S = (S_{p_i} + S_{q_i})/2$, and $|\cdot|$ denotes the matrix determinant. The first summand coincides with Hotelling's T^2 statistic except for minor differences in the weighting of the involved terms. Consequently, using an analogous approach in terms of the centralized covariance $\widetilde{\Sigma}_{(\cdot)}$ provides a consistent and bi-invariant generalization. Indeed, the second summand is also bi-invariant. To verify this, let $(g_1, \ldots, g_m)$ be a data set in a Lie group G with bi-invariant mean $\overline{g}$. For any group element $f \in G$, we have that $\log((f\overline{g})^{-1}(fg_i)) = \log(\overline{g}^{-1}g_i)$ and, thus, $\widetilde{\Sigma}_{g_i}$ left invariant. For right invariance, we can take advantage of the relationship $\log(fgf^{-1}) = \mathrm{Ad}(f)\log(g)$ [19, Thm. 6], yielding $\log((\overline{g}f)^{-1}(g_if)) = \mathrm{Ad}(f^{-1})\log(\overline{g}^{-1}g_i)$ and, thus,

$$[\widetilde{\Sigma}_{g_if}] = [\mathrm{Ad}(f^{-1})][\widetilde{\Sigma}_{g_i}][\mathrm{Ad}(f^{-1})]^T.$$

Since $\mathrm{Ad}(f^{-1})$ is invertible, the determinant $\rho_f = |[\mathrm{Ad}(f^{-1})]|$ is non-zero and we obtain $|[\widetilde{\Sigma}_{g_if}]| = \rho_f^2|[\widetilde{\Sigma}_{g_i}]|$. A simple calculation shows that the scaling ρ_f^2 cancels in the second summand, thus, verifying right invariance.

4 Experiments

We evaluate the proposed group test for the morphometric analysis of pathological malformations associated to cognitive decline, viz. mild cognitive impairment (MCI). MCI in the elderly is a common condition and often represents an intermediate stage between normal cognition and Alzheimer's disease. As consistently reported in neuroimaging studies, atrophy of the hippocampal formation is a characteristic early sign of MCI. In this section, we analyze hippocampal atrophy patterns due to MCI by applying the derived Hotelling's T^2 statistic to infer significant differences.

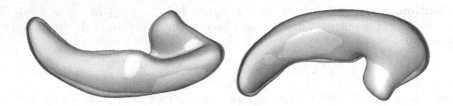

Fig. 1. Bi-invariant means of right hippocampi for cognitive normal (red, transparent) and impaired (white) subjects overlaid onto each other. (Color figure online)

4.1 Data Description

For our experiments we prepared a data set consisting of 26 subjects showing mild cognitive impairment (MCI) and 26 cognitive normal (CN) controls from the open access Alzheimer's Disease Neuroimaging Initiative[1] (ADNI) database. ADNI provides, among others, 1632 brain MRI scans collected on four different time points with segmented hippocampi. We established surface correspondence (2280 vertices, 4556 triangles) in a fully automatic manner employing the deblurring and denoising of functional maps approach [10] for isosurfaces extracted from the available segmentations. The dataset was randomly assembled from the baseline shapes for which segmentations were simply connected and remeshed surfaces were well-approximating ($\leq 10^{-5}$ mm root mean square surface distance to the isosurface).

4.2 GL$^+$(3)-Based Shape Space

For shape analysis we employ a recent representation [3] that describes shapes in terms of linear differential coordinates viewed as elements of GL$^+$(3). Given deformations $(\phi_1, \ldots, \phi_n)$ mapping a reference or template configuration $\bar{S}$ to surfaces $(S_1, \ldots, S_n)$, the coordinates—being the Jacobian matrices—provide a local characterization of the respective deformation and, thus, the shape changes. In particular, let ϕ_i be an orientation-preserving, simplicial map, then the derivatives are constant on each triangle T, viz. $\nabla \phi_i|_T \equiv D_i^T \in$ GL$^+$(3). Note, that the deformation of a triangle fully specifies an affine map of $\mathbb{R}^3$ assuming that triangle normals are mapped onto each other (cf. Kirchhoff–Love kinematic assumptions). Finally, obtaining a surface $\phi(\bar{S})$ for given coordinates leads to a linear differential equation that can be solved very efficiently.

4.3 Hippocampal Atrophy Patterns in CN vs. MCI

We compute bi-invariant means for the ADNI data set described in Sect. 4.1. A qualitative comparison is shown in Fig. 1 illustrating the well-known [15] loss of total hippocampal volume associated with MCI.

---

[1] adni.loni.usc.edu.

Fig. 2. Group test for differences between means of right hippocampi for cognitive normal and impaired subjects: p-values (FDR corrected) are colored coded using the colormap 0.0 ■■■■■■■ 0.05. (Color figure online)

Next, we evaluate the local differences in shape between the bi-invariant means by performing triangle-wise, partial tests that provide marginal information for each specific triangle allowing to investigate which subregions contribute significant differences. While Hotelling's T^2 statistic is based on quite stringent assumptions on the distribution, it can be utilized to derive a nonparametric testing procedure. In particular, we employ a permutation testing setup based on the proposed statistic (Definition 2) yielding a bi-invariant, distribution-free two-sample test. The key idea is to estimate the empirical distribution of the test statistic under the null-hypothesis H_0 that the two distributions to be tested are the same. To this end, group memberships of the observations are repeatedly permuted each time re-computing the statistic between the accordingly changed groups. The p-value is then computed as the proportion of test statistics that are greater than the one computed for the original (unpermuted) groups.

In Fig. 2 we visualize the regions with statistical significant differences ($p < 0.05$ after Benjamini-Hochberg false discovery correction) between the bi-invariant means showing the respective p-values. In line with literature on MCI [15], the obtained results suggest more differentiated morphometric changes beyond homogeneous volumetric decline of the hippocampi.

5   Discussion

In this work, we derived generalizations of established indices for the quantization of dissimilarity between empirically-defined probability distributions in Lie groups, viz. the Hotelling's T^2 statistic and the Bhattacharayya distance. These new measures are stable according to group operations (left/right composition and inversion), e.g. removing any bias due to arbitrary choices of a reference frame. Moreover, the generalizations are consistent to the definitions in multi-variate statistics, i.e. they agree for the special case of flat vector spaces. We further obtained nonparametric two-sample tests based on the proposed measures and validated them in group tests on malformations of right hippocampi due to mild cognitive impairment. While this experiment serves as an illustrating example, we plan to extend the analysis employing global and more strict simultaneous tests as, e.g., in [21].

As with other non-Euclidean approaches, the derived methods pose certain assumptions on the uniqueness and smeariness of the intrinsic mean [9]. Another assumption in the derivation of the Mahalanobis distances is the invertability of the covariance operator, which is frequently violated, e.g. when the number of observations is lower than the number of variables. A common approach in such situations is to resort to a pseudo-inverse (see e.g. [12]) of the covariance. Such a strategy, however, will not result in a bi-invariant notion of Mahalanobis distance. Extending the proposed expressions to such high dimension low sample size scenarios poses another interesting direction for future work.

Acknowledgments. M. Hanik is funded by the Deutsche Forschungsgemeinschaft (DFG, German Research Foundation) under Germany's Excellence Strategy - The Berlin Mathematics Research Center MATH+ (EXC-2046/1, project ID: 390685689). We are grateful for the open-access dataset of the Alzheimer's Disease Neuroimaging Initiative (ADNI) (Data collection and sharing for this project was funded by the ADNI (National Institutes of Health Grant U01 AG024904) and DOD ADNI (Department of Defense award number W81XWH-12-2-0012). ADNI is funded by the National Institute on Aging, the National Institute of Biomedical Imaging and Bioengineering, and through generous contributions from the following: AbbVie, Alzheimer's Association; Alzheimer's Drug Discovery Foundation; Araclon Biotech; BioClinica, Inc.; Biogen; Bristol-Myers Squibb Company; CereSpir, Inc.; Cogstate; Eisai Inc.; Elan Pharmaceuticals, Inc.; Eli Lilly and Company; EuroImmun; F. Hoffmann-La Roche Ltd and its affiliated company Genentech, Inc.; Fujirebio; GE Healthcare; IXICO Ltd.; Janssen Alzheimer Immunotherapy Research & Development, LLC.; Johnson & Johnson Pharmaceutical Research & Development LLC.; Lumosity; Lundbeck; Merck & Co., Inc.; Meso Scale Diagnostics, LLC.; NeuroRx Research; Neurotrack Technologies; Novartis Pharmaceuticals Corporation; Pfizer Inc.; Piramal Imaging; Servier; Takeda Pharmaceutical Company; and Transition Therapeutics. The Canadian Institutes of Health Research is providing funds to support ADNI clinical sites in Canada. Private sector contributions are facilitated by the Foundation for the National Institutes of Health (www.fnih.org). The grantee organization is the Northern California Institute for Research and Education, and the study is coordinated by the Alzheimer's Therapeutic Research Institute at the University of Southern California. ADNI data are disseminated by the Laboratory for Neuro Imaging at the University of Southern California.) as well as for F. Ambellan's help in establishing dense correspondences of the hippocampal surface meshes.

References

1. Adler, R.L., Dedieu, J., Margulies, J.Y., Martens, M., Shub, M.: Newton's method on Riemannian manifolds and a geometric model for the human spine. IMA J. Numer. Anal. **22**(3), 359–390 (2002). https://doi.org/10.1093/imanum/22.3.359
2. Ambellan, F., Lamecker, H., von Tycowicz, C., Zachow, S.: Statistical shape models: understanding and mastering variation in anatomy. In: Rea, P.M. (ed.) Biomedical Visualisation. AEMB, vol. 1156, 1st edn., pp. 67–84. Springer, Cham (2019). https://doi.org/10.1007/978-3-030-19385-0_5

3. Ambellan, F., Zachow, S., von Tycowicz, C.: An as-invariant-as-possible GL$^+$(3)-based statistical shape model. In: Zhu, D., et al. (eds.) MBIA/MFCA -2019. LNCS, vol. 11846, pp. 219–228. Springer, Cham (2019). https://doi.org/10.1007/978-3-030-33226-6_23

4. Ambellan, F., Zachow, S., von Tycowicz, C.: A surface-theoretic approach for statistical shape modeling. In: Shen, D., et al. (eds.) MICCAI 2019. LNCS, vol. 11767, pp. 21–29. Springer, Cham (2019). https://doi.org/10.1007/978-3-030-32251-9_3

5. Bhattacharyya, A.: On a measure of divergence between two multinomial populations. Sankhyā: Indian J. Stat. **7**, 401–406 (1946)

6. Boisvert, J., Cheriet, F., Pennec, X., Labelle, H., Ayache, N.: Geometric variability of the scoliotic spine using statistics on articulated shape models. IEEE Trans. Med. Imaging **27**(4), 557–568 (2008). https://doi.org/10.1109/TMI.2007.911474

7. do Carmo, M.P.: Riemannian Geometry. Mathematics: Theory and Applications, 2nd edn. Birkhäuser, Boston (1992)

8. Cartan, E., Shouten, J.: On the geometry of the group-manifold of simple and semi-groups. Proc. Akad. Wetensch. Amsterdam **29**, 803–815 (1926)

9. Eltzner, B., Huckemann, S.F.: A smeary central limit theorem for manifolds with application to high-dimensional spheres. Ann. Stat. **47**(6), 3360–3381 (2019). https://doi.org/10.1214/18-AOS1781

10. Ezuz, D., Ben-Chen, M.: Deblurring and denoising of maps between shapes. Comput. Graph. Forum **36**, 165–174 (2017). https://doi.org/10.1111/cgf.13254. Wiley Online Library

11. Grenander, U.: General Pattern Theory: A Mathematical Study of Regular Structures. Oxford Mathematical Monographs. Clarendon Press, Oxford (1993)

12. Hong, Y., Singh, N., Kwitt, R., Niethammer, M.: Group testing for longitudinal data. In: Ourselin, S., Alexander, D.C., Westin, C.-F., Cardoso, M.J. (eds.) IPMI 2015. LNCS, vol. 9123, pp. 139–151. Springer, Cham (2015). https://doi.org/10.1007/978-3-319-19992-4_11

13. Van Hecke, W., Emsell, L., Sunaert, S. (eds.): Diffusion Tensor Imaging. Springer, New York (2016). https://doi.org/10.1007/978-1-4939-3118-7

14. Miller, M., Younes, L.: Group actions, homeomorphisms, and matching: a general framework. Int. J. Comput. Vision **41**, 61–84 (2001). https://doi.org/10.1023/A:1011161132514

15. Mueller, S.G., et al.: Hippocampal atrophy patterns in mild cognitive impairment and Alzheimer's disease. Hum. Brain Mapp. **31**(9), 1339–1347 (2010). https://doi.org/10.1002/hbm.20934

16. Muralidharan, P., Fletcher, P.: Sasaki metrics for analysis of longitudinal data on manifolds. In: Proceedings of the 2012 IEEE Conference on Computer Vision and Pattern Recognition (CVPR), vol. 2012, pp. 1027–1034 (2012). https://doi.org/10.1109/CVPR.2012.6247780

17. Pennec, X., Sommer, S., Fletcher, T.: Riemannian Geometric Statistics in Medical Image Analysis. Elsevier Science & Technology, Amsterdam (2019). https://doi.org/10.1016/C2017-0-01561-6

18. Pennec, X.: Intrinsic statistics on Riemannian manifolds: basic tools for geometric measurements. J. Math. Imaging Vis. **25**, 127–154 (2006). https://doi.org/10.1007/s10851-006-6228-4

19. Pennec, X., Arsigny, V.: Exponential barycenters of the canonical Cartan connection and invariant means on lie groups. In: Nielsen, F., Bhatia, R. (eds.) Matrix Information Geometry, pp. 123–166. Springer, Heidelberg (2013). https://doi.org/10.1007/978-3-642-30232-9_7

20. Postnikov, M.: Geometry VI: Riemannian Geometry. Encyclopaedia of Mathematical Sciences. Springer, Heidelberg (2013). https://doi.org/10.1007/978-3-662-04433-9
21. Schulz, J., Pizer, S., Marron, J., Godtliebsen, F.: Non-linear hypothesis testing of geometric object properties of shapes applied to hippocampi. J. Math. Imaging Vis. **54**, 15–34 (2016). https://doi.org/10.1007/s10851-015-0587-7
22. Thompson, D.W.: On Growth and Form. Canto. Cambridge University Press, Cambridge (1992). https://doi.org/10.1017/cbo9781107325852
23. von Tycowicz, C., Ambellan, F., Mukhopadhyay, A., Zachow, S.: An efficient Riemannian statistical shape model using differential coordinates. Med. Image Anal. **43**, 1–9 (2018). https://doi.org/10.1016/j.media.2017.09.004
24. Woods, R.P.: Characterizing volume and surface deformations in an atlas framework: theory, applications, and implementation. NeuroImage **18**(3), 769–788 (2003). https://doi.org/10.1016/s1053-8119(03)00019-3

Learning

Uncertain-DeepSSM: From Images to Probabilistic Shape Models

Jadie Adams[1,2(✉)], Riddhish Bhalodia[1,2(✉)], and Shireen Elhabian[1,2(✉)]

[1] Scientific Computing and Imaging Institute, University of Utah,
Salt Lake City, UT, USA
{jadie,riddhishb,shireen}@sci.utah.edu
[2] School of Computing, University of Utah, Salt Lake City, UT, USA

Abstract. Statistical shape modeling (SSM) has recently taken advantage of advances in deep learning to alleviate the need for a time-consuming and expert-driven workflow of anatomy segmentation, shape registration, and the optimization of population-level shape representations. DeepSSM is an end-to-end deep learning approach that extracts statistical shape representation directly from unsegmented images with little manual overhead. It performs comparably with state-of-the-art shape modeling methods for estimating morphologies that are viable for subsequent downstream tasks. Nonetheless, DeepSSM produces an overconfident estimate of shape that cannot be blindly assumed to be accurate. Hence, conveying what DeepSSM does not know, via quantifying granular estimates of uncertainty, is critical for its direct clinical application as an on-demand diagnostic tool to determine how trustworthy the model output is. Here, we propose Uncertain-DeepSSM as a unified model that quantifies both, data-dependent aleatoric uncertainty by adapting the network to predict intrinsic input variance, and model-dependent epistemic uncertainty via a Monte Carlo dropout sampling to approximate a variational distribution over the network parameters. Experiments show an accuracy improvement over DeepSSM while maintaining the same benefits of being end-to-end with little pre-processing.

Keywords: Uncertainty quantification · Statistical shape modeling · Bayesian deep learning

1 Introduction

Morphometrics and its new generation, statistical shape modeling (SSM), have evolved into an indispensable tool in medical and biological sciences to study anatomical forms. SSM has enabled a wide range of biomedical and clinical applications (e.g., [2,4,6,7,17–19,22,31,47,49]). Morphological analysis requires parsing the anatomy into a quantitative representation consistent across the population at hand to facilitate the testing of biologically relevant hypotheses. A popular choice for such a representation is using *landmarks* that are defined consistently using invariant points, i.e., *correspondences*, across populations [43].

© Springer Nature Switzerland AG 2020
M. Reuter et al. (Eds.): ShapeMI 2020, LNCS 12474, pp. 57–72, 2020.
https://doi.org/10.1007/978-3-030-61056-2_5

Coordinate transformations (e.g., [26,27]) hold promise as an alternative representation, but the challenge is finding the anatomically-relevant transformation that quantifies differences among shapes. Ideally, landmarking is performed by anatomy experts to mark distinct, and typically few anatomical features [1,33], but it is time-intensive and cost-prohibitive, especially for 3D images and large cohorts. More recently, dense sets of correspondence points that capture population statistics are used, thanks to advances in computationally driven approaches for shape modeling (e.g., [8,9,11,13,45]).

Traditional computational approaches to automatically generate dense correspondence models, a.k.a. point distribution models (PDMs), still entail a time-consuming, expert-driven, and error-prone workflow of segmenting anatomies from volumetric images, followed by a processing pipeline of shape registration, correspondence optimization, and projecting points onto some low-dimensional shape space for subsequent statistical analysis. Many of these steps require significant parameter tuning and/or quality control by the users. The excessive time and effort to construct population-specific shape models have motivated the use of deep networks and their inherent ability to learn complex functional mappings to regress shape information directly from images and incorporate prior knowledge of shapes in image segmentation tasks (e.g., [3,25,38,48,50]). However, deep learning in this context has drawbacks. Training deep networks on volumetric images is often confounded by the combination of high-dimensional image spaces and limited availability of training images labeled with shape information. Additionally, deep networks can make poor predictions with no indication of uncertainty when the training data weakly represents the input. Computationally efficient automated morphology assessment when integrated with new clinical tools as well as surgical procedure, has potential to improve medical

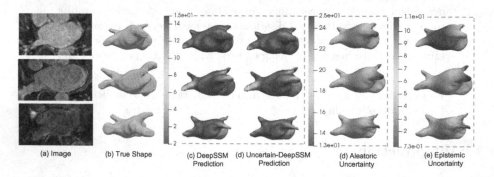

(a) Image   (b) True Shape   (c) DeepSSM (d) Uncertain-DeepSSM   (d) Aleatoric   (e) Epistemic
                              Prediction   Prediction   Uncertainty   Uncertainty

Fig. 1. Shape prediction and uncertainty quantification on left atrium MRI scans. The images (a) are input and the true shapes (surface meshes) (b) are from ground truth segmentations. Shapes in (c) and (d) are constructed from DeepSSM and Uncertain-DeepSSM predictions, respectively. The heat maps on surface meshes in (c) and (d) show the surface-to-surface distance to (b) (the error in mm). The aleatoric (d) and epistemic (e) output from Uncertain-DeepSSM are shown as heat maps on the predicted mesh. Our model outputs increased uncertainty where error is high.

care standards and clinical decision making. However, uncertainty quantification is a must in such scenarios, as it will allow professionals to determine the trustworthiness of such a tool and prevent unsafe predictions.

Here, we focus on a particular instance of a deep learning-based framework, namely DeepSSM [3], that maps unsegmented 3D images to a low-dimensional shape descriptor. Mapping to a low-dimensional manifold, compared with regressing correspondence points, has a regularization effect that compensates for misleading image information and provides a topology-preserving prior to shape estimation. DeepSSM also entails a population-driven data augmentation approach that addresses limited training data, which is typical with small and large-scale shape variability. DeepSSM has been proven effective in characterizing pathology [3] and performs statistically similar to traditional SSM methods in downstream tasks such as disease recurrence predictions [4]. Nonetheless, DeepSSM, like other deep learning-based frameworks predicting shape, produces an overconfident estimate of shape that can not be blindly assumed to be accurate. Furthermore, the statistic-preserving data augmentation is bounded by what the finite set of training samples captures about the underlying data distribution. In this paper, we formalize Uncertain-DeepSSM, a unified solution to limited training images with dense correspondences and model prediction over-confidence. Uncertain-DeepSSM quantifies granular estimates of uncertainty with respect to a low-dimensional shape descriptor to provide spatially-coherent, localized uncertainty measures (see Fig. 1) that are robust to misconstruing factors that would typically affect point-wise regression, such as heterogeneous image intensities and noisy diffuse shape boundaries. Uncertain-DeepSSM produces probabilistic shape models directly from 3D images, conveying what DeepSSM does not know about the input and providing an accuracy improvement over DeepSSM while being end-to-end with little required pre-processing.

2 Related Work

DeepSSM [3] is based on works that show the efficacy of convolutional neural networks (CNNs) to extract shape information from images. Huang *et al.* [25] regress shape orientation and position conditioned on 2D ultrasound images. Milletari *et al.* [38] segment ultrasound images using low-dimensional shape representation in the form of principal component analysis (PCA) to regress landmark positions. Oktay *et al.* [41] incorporate prior knowledge about organ shape and location into a deep network to anatomically constrain resulting segmentations. However, these works provide a point-estimate solution for the task at hand.

In Bayesian modeling, there are two types of uncertainties [28,30]. *Aleatoric* (or data) uncertainty captures the uncertainty inherent in the input data, such as over-exposure, noise, and the lack of the image-based features indicative of shapes. *Epistemic* (or model) uncertainty accounts for uncertainty in the model parameters and can be explained away, given enough training data [28].

Aleatoric uncertainty can be captured by placing a distribution over the model output. In image segmentation tasks, this has been achieved by sampling segmentations from an estimated posterior distribution [10,34] and using conditional normalizing flows [44] to infer a distribution of plausible segmentations conditioned on the input image. These efforts succeed in providing shape segmentation with aleatoric uncertainty measures, but do not provide a shape representation that can be readily used for population-level statistical analyses. Tóthová *et al.* [46] incorporate prior shape information into a deep network in the form of a PCA model to reconstruct surfaces from 2D images with an aleatoric uncertainty measure that is quantified via conditional probability estimation. Besides being limited to 2D images, quantifying point-wise aleatoric uncertainty makes this measure prone to inherent noise in images.

Epistemic uncertainty is more difficult to model as it requires placing distributions over models rather than their output. Bayesian neural networks [12,14,37] achieve this by placing a prior over the model parameters, then quantifying their variability. Monte Carlo dropout sampling, which places a Bernoulli distribution over model parameters [15], has effectively been formalized as a Bayesian approximation for capturing epistemic uncertainty [16]. Aleatoric and epistemic uncertainty measures have been combined in one model for tasks such as semantic segmentation, depth regression, classification, and image translation [28,32,42], but never for SSM.

Uncertain-DeepSSM produces probabilistic shape models directly from images that quantifies both the data-dependent aleatoric uncertainty and the model-dependent epistemic uncertainty. We quantify aleatoric uncertainty by adapting the network to predict intrinsic input variance in the form of mean and variance for the PCA scores and updating the loss function accordingly [35,40]. This enables explicit modeling of the heteroscedastic-type of aleatoric uncertainty, which is dependent on the input data. We model epistemic uncertainty via a Monte Carlo dropout sampling to approximate a variational distribution over the network parameters by sampling PCA score predictions with various dropout masks. This approach provides both uncertainty measures for each PCA score that are then mapped back to the shape space for interpretable visualization. Uncertainty fields on estimated 3D shapes convey insights for how the given input relate to what the model knows. For instance, such uncertainties could help pre-screen for pathology if Uncertain-DeepSSM is trained on controls. Furthermore, explicit modeling of uncertainties in Uncertain-DeepSSM provides more accurate predictions, compared with DeepSSM, with no additional training steps. This indicates the ability of Uncertain-DeepSSM to better generalize in limited training data setting.

3   Methods

A trained Uncertain-DeepSSM model provides shape descriptors, specifically PCA scores, with uncertainty measures directly from 3D images (e.g., CT, MRI) of anatomies. In this section, we describe the data augmentation method, the network architecture, training strategy, and uncertainty quantification.

3.1   Notations

Consider a paired dataset $\{(\mathbf{x}_n, \mathbf{y}_n)\}_{n=1}^N$ of N 3D images $\mathbf{y}_n \in \mathbb{R}^{H \times W \times D}$ and their corresponding shapes $\mathbf{x}_n \in \mathbb{R}^{3M}$, where each shape is represented by M 3D correspondence points. We generate a PDM from segmentations. This entails the typical SSM pipeline that includes pre-processing steps (registration, resampling, smoothing, ...), and correspondence (i.e., PDM) optimization. In practice, any PDM generation algorithm can be employed. Here, we use the open-source *ShapeWorks* software [8] to optimize surface correspondences using anatomies segmented from the *training* images. Next, high-dimensional shapes (i.e., PDM) in the shape space (of dimension $\mathbb{R}^{3M}$) are mapped to low-dimensional PCA scores $\mathbf{z} \in \mathbb{R}^L$ in the PCA subspace that is parameterized by a mean vector $\boldsymbol{\mu} \in \mathbb{R}^{3M}$, a diagonal matrix of eigen values $\boldsymbol{\Delta} \in \mathbb{R}^{L \times L}$, and a matrix of eigen vectors $\mathbf{U} \in \mathbb{R}^{3M \times L}$, where $\mathbf{z} = \mathbf{U}^T(\mathbf{x} - \boldsymbol{\mu})$ and $L \ll 3M$ is chosen such that at least 95% of the population variation is explained. The PCA scores $\mathbf{z}_n$ associated with the training image $\mathbf{y}_n$ serve as a supervised target to be inferred by the Uncertain-DeepSSM network and mapped deterministically to correspondence points, where $\mathbf{x}_n = \mathbf{U}\mathbf{z}_n + \boldsymbol{\mu}$. The network thus defines a functional map $f_{\boldsymbol{\Theta}} : \mathbb{R}^{H \times W \times D} \to \mathbb{R}^L$ that is parameterized by the network parameters $\boldsymbol{\Theta}$, where $\mathbf{z} = f_{\boldsymbol{\Theta}}(\mathbf{y})$. Uncertainties are quantified in the PCA subspace, such that the PCA scores of the n–th training shape $\mathbf{z}_n$ is associated with vectors of aleatoric variances $\mathbf{a}_n \in \mathbb{R}_+^L$ and epistemic variances $\mathbf{e}_n \in \mathbb{R}_+^L$.

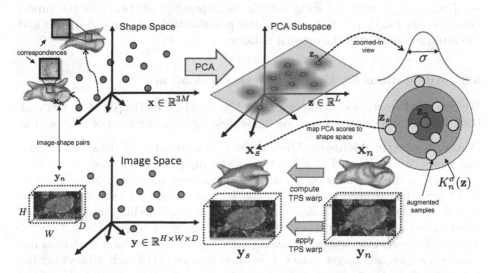

Fig. 2. Data augmentation: PCA is used to compute the PCA scores $\{\mathbf{z}_n\}_{n=1}^N$ of the training shape samples $\{\mathbf{x}_n\}_{n=1}^N$. Augmented samples $\mathbf{z}_s$ are randomly drawn from a collection of multivariate Gaussian distributions $K_\sigma^n(\mathbf{z})$, each with a training example $\mathbf{z}_n$ as the mean and covariance $\sigma^2 \mathbf{I}_L$. The correspondences $\mathbf{x}_s$ are used to compute a TPS warp to map $\mathbf{x}_n$ to $\mathbf{x}_s$, which is used to warp the respective image $\mathbf{y}_n$ to a new image $\mathbf{y}_s$ with known shape parameters $\mathbf{z}_s$.

3.2   Data Augmentation

DeepSSM augments training data with shape samples generated from a single multivariate Gaussian distribution in the PCA subspace, where an add-reject strategy is employed to prevent outliers from being sampled. Instead of assuming a Gaussian distribution, we use a kernel density estimate (KDE) to better capture the underlying shape distribution by not over-estimating the variance and avoiding sampling implausible shapes from high probability regions in case of multi-modal distributions. Using KDE, augmented samples $\mathbf{z}_s$ are drawn from:

$$p_\sigma(\mathbf{z}) = \frac{1}{N} \sum_{n=1}^{N} K_n^\sigma(\mathbf{z}), \quad \text{s.t.} \quad K_n^\sigma(\mathbf{z}) = \frac{1}{(2\pi\sigma^2)^{L/2}} \exp\left(-\frac{||\mathbf{z} - \mathbf{z}_n||^2}{2\sigma^2}\right), \quad (1)$$

where $\sigma \in \mathbb{R}_+$ denotes the kernel bandwidth and is computed as the average nearest neighbor distance in the PCA subspace, i.e., $\sigma^2 = \frac{1}{N}\sum_n \min_{k \neq n}(\mathbf{z}_n - \mathbf{z}_k)^T \mathbf{\Delta}^{-1}(\mathbf{z}_n - \mathbf{z}_k)$. As illustrated in Fig. 2, a sampled vector of PCA scores $\mathbf{z}_s \in \mathbb{R}^L$ from the kernel of the n−th training sample $K_n^\sigma(\mathbf{z})$ is mapped to correspondence points $\mathbf{x}_s \in \mathbb{R}^{3M}$, where $\mathbf{x}_s = \mathbf{U}\mathbf{z}_s + \boldsymbol{\mu}$. Using the $\mathbf{x}_n \leftrightarrow \mathbf{x}_s$ correspondences, we compute thin-plate spline (TPS) warp [5] to obtain a deformation field which is then applied to image $\mathbf{y}_n$ to construct the augmented image $\mathbf{y}_s$. With this augmentation method, we can construct an augmented training set $\{(\mathbf{x}_s, \mathbf{y}_s)\}_{s=1}^S$ of S 3D images, their corresponding shapes, and the supervised targets $\{\mathbf{z}_s\}_{s=1}^S$, which respects the population-level shape statistics and the intensity profiles of the original dataset.

3.3   Adaptations for Uncertainty Quantification

We extend the network architecture and loss function of DeepSSM to estimate both types of uncertainties and the shape descriptor in the form of PCA scores.

Heteroscedastic Aleatoric Uncertainty is a measure of data uncertainty, and hence can be learned as a function of the input. Given a training set $\mathcal{D} = \{(\mathbf{y}_i, \mathbf{z}_i)\}_{i=1}^I$ that includes both real and augmented samples, where $I = N + S$, DeepSSM is trained to minimize the L2 loss between groundtruth $\mathbf{z}_i$ and predicted $f_\Theta(\mathbf{y}_i)$. In Uncertain-DeepSSM, the network architecture is modified to estimate both the mean $\bar{\mathbf{z}}_i$ and variance $\mathbf{a}_i$ of the PCA scores, where $[\bar{\mathbf{z}}_i, \mathbf{a}_i] = f_\Theta(\mathbf{y}_i)$. The variance acts as an uncertainty regularization term that does not require a supervised target since it is learned implicitly through supervising the regression task. For training purposes, we let the network predict the log of the variance, $\tilde{a}_{il} = \log a_{il}^2$, where a_{il} captures the aleatoric uncertainty along the l−th PCA mode of variation. This forces the variance to be positive and removes the potential for division by zero. Uncertain-DeepSSM is thus trained to minimize the Bayesian loss in (2), where $\bar{\mathbf{z}}_i = f_\Theta^{\mathbf{z}}(\mathbf{y}_i)$ and $\tilde{\mathbf{a}}_i = f_\Theta^{\mathbf{a}}(\mathbf{y}_i)$ are the $\mathbf{z}$− and $\mathbf{a}$− outputs of the network, respectively (see Fig. 3):

$$\mathcal{L}(\boldsymbol{\Theta}) = \frac{1}{2LI} \sum_{i=1}^{I} \left\{ [\mathbf{z}_i - \bar{\mathbf{z}}_i]^T \operatorname{diag}\left(\exp(\widetilde{\mathbf{a}}_i)\right)^{-1} [\mathbf{z}_i - \bar{\mathbf{z}}_i] + \sum_{l=1}^{L} \widetilde{a}_{il} \right\}. \tag{2}$$

The second term in (2) learns a loss attenuation, preventing the network from predicting infinite variance for all scores.

Epistemic uncertainty is a measure of the model's ignorance that can be quantified by modeling distributions over the model parameters $\boldsymbol{\Theta}$. We place a Bernoulli distribution over network weights by making use of the Monte Carlo dropout sampling technique [15,16]. In particular, a dropout layer with a probability κ is added to every layer (convolutional and fully connected) in the Uncertain-DeepSSM network (Fig. 3). Dropout is used in both, training and testing, where in testing, it is used to sample from the approximate posterior. The various dropout masks provide an ensemble of networks to sample predictions from, the distribution of which reflects the model's epistemic uncertainty. Consider V dropout samples, the epistemic uncertainty of the l-th PCA mode of variation is computed as,

$$e_{il} = \frac{1}{V} \sum_{v=1}^{V} \left(\bar{z}_{il}^{(v)} \right)^2 - \left(\frac{1}{V} \sum_{v=1}^{V} \bar{z}_{il}^{(v)} \right)^2 \tag{3}$$

where $\bar{\mathbf{z}}_i^{(v)} = f_{\boldsymbol{\Theta}_v \sim p(\boldsymbol{\Theta})}^{\mathbf{z}}(\mathbf{y}_i)$ is the $\mathbf{z}$-output of the network for the randomly masked network parameters $\boldsymbol{\Theta}_v$.

3.4   Architecture and Training

The network architecture of Uncertain-DeepSSM (Fig. 3) is similar to DeepSSM with five convolution layers followed by two fully connected layers. However in Uncertain-DeepSSM, dropout is added and batch normalization is removed. Combining batch normalization and dropout leads to a variance shift that causes training instability [36]. Hence, we normalize the input images to compensate for not using batch normalization. The PCA scores are also whitened to prevent the model from favoring the dominant PCA modes. A dropout layer with a probability of $\kappa = 0.2$ is added after every convolutional and fully connected layer. Data augmentation is used to create a set of $I = 4000$ training and 1000 validation images for training the network. PyTorch is used in constructing and training DeepSSM with Adam optimization [29] and a learning rate of 0.0001. Parametric ReLU [23] activation and Xavier weight initialization [21] are used. To train Uncertain-DeepSSM, the L2 loss function is used for the first epoch and the Bayesian loss function (2) is used for all following epochs. This allows the network to learn based on the task alone before learning to quantify uncertainty, resulting in better predictions and more stable training.

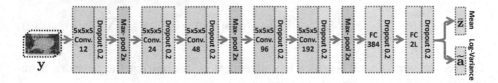

Fig. 3. Uncertain-DeepSSM network architecture

3.5   Testing and Uncertainty Analysis

When testing, dropout remains on and predictions are sampled multiple times. The predicted PCA scores and aleatoric uncertainty measure are first un-whitened. Using the dropout samples, the epistemic uncertainty measure is computed using (3) based on the un-whitened predicted scores. To compute the accuracy of the predictions, we first map PCA scores to correspondence points and compare the surface reconstructed from these points to the surface constructed from the ground truth segmentation. For surface reconstruction, we use the point correspondences between the population mean and the correspondence points of the predicted PCA scores to define a TPS warp that deform the mean mesh (from ShapeWorks [8]) to obtain the surface mesh for the predicted scores. The error is then calculated as the average of the surface-to-surface distance from the predicted to ground truth mesh and that of ground truth to predicted mesh.

To visualize uncertainty measures on the predicted mesh, the location of each correspondence point is modeled as a Gaussian distribution. To fit these distributions, we sample PCA scores from a Gaussian with the predicted mean and desired variance (aleatoric or epistemic), then map them to the PDM space. This provides us with a distribution over each correspondence point, with mean and entropy indicating the coordinates of the point and the associated uncertainty scalar, respectively. Interpolation is then used to interpolate uncertainty scalars defined on correspondence points to the full reconstructed mesh.

4   Results

We compare the shape predictions from DeepSSM and Uncertain-DeepSSM on two 3D datasets; a toy dataset of parametric shapes (supershapes) as a proof-of-concept and a real world dataset of left atrium MRI scans. In both experiments, we create three different test sets: *control*, *aleatoric*, and *epistemic*. The control test set is well represented under the training population, whereas the epistemic and aleatoric test sets are not. Examples with images that differ from the training images are chosen for the aleatoric set, as this suggests data uncertainty. The epistemic set is chosen to demonstrate model uncertainty by selecting examples with shapes that differ from those in the training set. The test sets are held out from the entire data augmentation and training process. It is important that test sets are not used to build the PDMs, such that they are not reflected in the population statistics captured in the PCA scores. Hence, we use surface-to-surface distances between meshes to quantify shape-based prediction errors since testing samples do not have optimized (ground truth) correspondences.

4.1 Supershapes Dataset

As a proof-of-concept, we construct a set of 3D supershapes shapes [20], which are a family of parameterized shapes. A supershape is parameterized by three variables, one which determines the number of lobes in the shape (or the shapes group), and two which determine the curvature of the shape. To create the training and validation sets, we generate 5000 3-lobe shapes with randomly drawn curvature values (using a χ^2 distribution). For each shape, a corresponding image of size $98 \times 98 \times 98$ is formed, where the intensities of the foreground and background are modeled as Gaussian distributions with different means but same variance. Additive Gaussian noise is added and the images are blurred with a Gaussian filter to mimic diffuse shape boundaries. In this case, Uncertain-DeepSSM predicts a single PCA score, where the first dominant PCA mode captures 99% of the shape variability.

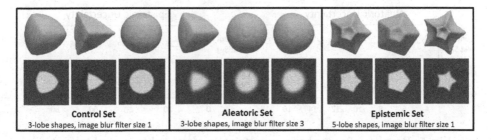

Fig. 4. Examples of surfaces and corresponding image slices from supershapes test sets.

We analyze output uncertainty measures on three different test sets, each of size 100. Examples of these can be seen in Fig. 4. The control test set is generated in the same manner as the training data and provides baseline uncertainty measures. The aleatoric test set contains shapes of the same shape group as the training, but the corresponding images are blurred with a larger Gaussian filter. This makes the shape boundary less clear, which has the effect of adding data uncertainty. For the epistemic test set, the images are blurred to the same degree as the training set, but the shapes belong to a different shape group. Here, we use 5-lobe shapes instead of 3-lobe to demonstrate model uncertainty.

The results of all three test sets are shown in Table 1. The predictions of Uncertain-DeepSSM are more accurate than DeepSSM on all of the test sets, with the aleatoric set having a notable difference. This is a result of the averaging effect of prediction sampling, which counters the effect of image blurring. The box plots of the uncertainty measure associated with the predicted PCA

Table 1. Average error and uncertainty measures on supershapes test sets.

	DeepSSM	Uncertain-DeepSSM		
	Surface-to-Surface distance	Surface-to-Surface distance	Aleatoric uncertainty	Epistemic uncertainty
Control test set	0.670 ± 0.104	0.615 ± 0.163	7.413 ± 2.189	15.000 ± 11.762
Aleatoric test set	1.293 ± 0.679	0.798 ± 0.447	10.205 ± 2.276	22.178 ± 13.065
Epistemic test set	7.045 ± 1.653	7.008 ± 1.668	12.256 ± 5.424	36.226 ± 17.327

score in Fig. 5 demonstrate that as expected, Uncertain-DeepSSM predicts higher aleatoric uncertainty on the aleatoric test set and higher epistemic uncertainty on the epistemic test set when compared to the control. The epistemic test has the highest of both forms of uncertainty because changing the shape group produces a great shift in the image domain (aleatoric) and shape domain (epistemic).

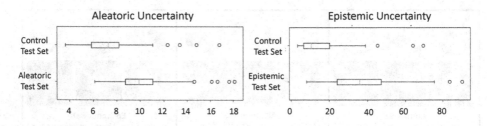

Fig. 5. Boxplots of supershapes uncertainties compared to control test set.

4.2 Left Atrium (LA) Dataset

The LA dataset consists of 206 late gadolinium enhancement MRI images of size $235 \times 138 \times 175$ that vary significantly in intensity and quality and have surrounding anatomical structures with similar intensities. The LA shape variation is also significant due to the topological variants pertaining to pulmonary veins arrangements [24]. The variation in images and shapes suggest a strong need for uncertainty measures. For networks training purposes, the images are downsampled to size $118 \times 69 \times 88$. We predict 19 PCA scores such that 95% of the shape-population variability is preserved. We compare DeepSSM and Uncertain-DeepSSM on three test sets, each of size 30. To define the aleatoric test set, we run PCA (preserving 95% of variability) on all 206 images. We then consider the Mahalanobis distance of the PCA scores of each sample to the mean PCA scores (within-subspace distance) as well as the image reconstruction error (mean

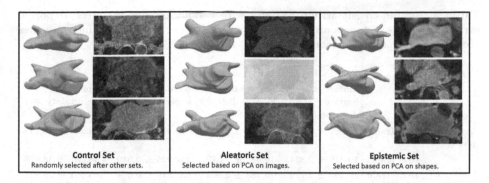

Fig. 6. Examples from of surfaces and image slices for LA test sets.

square error as off-subspace distance). These values are normalized and summed to get a measure of image similarity to the whole set (similar to [39]). We select the 30 that differ the most to be the aleatoric test set. These examples are the least supported by the input data, suggesting they should have high data uncertainty. To define the epistemic test set, we use the same technique but perform PCA on the signed distance transforms, as an implicit form of shapes, rather than the raw images. In this way, we are able to select an epistemic test set of 30 examples with shapes that are poorly supported by the data. This selection technique produces aleatoric and epistemic test sets that overlap by 6 examples, leaving 152 out of 206 samples. 30 of these are randomly selected to be the control test set and the rest (122) are used in data augmentation to create a training set of 4000 and validation set of 1000. Examples from the test sets can be seen in Fig. 6.

We train both DeepSSM and Uncertain-DeepSSM on different percentages of training data, namely 100%, 75%, and 25%, where an X% is randomly drawn from the remaining 122 samples and then used to proportionally augment the data. The average results of these tests are shown in Table 2. As expected, epistemic uncertainty measures decrease with more training data because model uncertainty can be explained away given more data. Uncertain-DeepSSM made more accurate predictions in most cases, notably when training data is limited. Uncertain-DeepSSM also successfully quantified uncertainty as we can see in the box plots in Fig. 7, which illustrate increased uncertainty measures on the uncertain test sets as compared to the control. The scatter plot in Fig. 7 illustrates the correlation between accuracy and uncertainty measures. The trend lines (combined based on all three test sets) indicate that the uncertainty quantification from Uncertain-DeepSSM provides insight into how trustworthy the model output is.

In Fig. 1, the uncertainty measures are shown on the meshes constructed from model predictions (models trained on 100% of the training data). The top example is from the control set, the second is from the aleatoric set, and the bottom is from the epistemic set. Here, we can see that both aleatoric and

Table 2. Results on left atrium test sets with various training set sizes. Reported surface-to-surface distances are averaged across the test set and uncertainty measures are averaged across PCA modes and the test set.

		DeepSSM	Uncertain-DeepSSM		
		Surface-to-Surface distance (mm)	Surface-to-Surface distance (mm)	Aleatoric uncertainty	Epistemic uncertainty
Control test set	25% Train	15.262 ± 3.694	10.670 ± 2.560	519.026 ±7.357	58.206 ± 38.145
	75% Train	10.319 ± 2.834	10.072 ± 2.812	452.025 ± 0.519	46.581 ± 32.678
	100% Train	10.205 ± 2.779	10.153 ± 2.904	431.518 ± 0.674	43.561 ± 29.821
Aleatoric test set	25% Train	12.967 ± 3.592	12.830 ± 3.543	472.359 ± 10.164	64.312 ± 45.656
	75% Train	12.507 ± 3.522	12.169 ± 3.493	465.951 ± 1.089	60.458 ± 45.454
	100% Train	12.242 ± 3.602	12.289 ± 3.525	442.129 ± 0.917	56.009 ± 42.371
Epistemic test set	25% Train	15.759 ± 4.301	14.975 ± 4.209	465.817 ± 7.567	75.854 ± 51.188
	75% Train	14.690 ± 4.166	14.581 ± 4.104	446.641 ± 1.127	64.236 ± 44.642
	100% Train	14.558 ± 4.151	14.465 ± 4.092	448.082 ± 1.291	61.517 ± 42.259

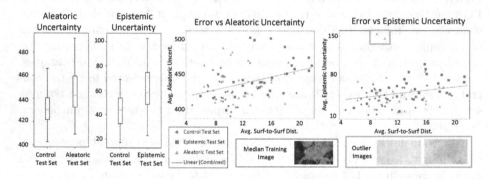

Fig. 7. Results on LA test sets from training Uncertain-DeepSSM on the full training set. The box plots show output uncertainty measures compared to the control test set. The scatter plot shows the average error versus average uncertainty on all three test sets. The two outliers marked with an orange box in the epistemic uncertainty plot are examples with images of a much higher intensity than the training examples (shown to the right) causing a spike in epistemic uncertainty.

epistemic uncertainty are higher in regions where the surface-to-surface distance is higher. This demonstrates the practicality of Uncertain-DeepSSM in a clinical setting as it indicates what regions of the predicted shape professionals can trust and where they should be skeptical.

5   Conclusion

Uncertain-DeepSSM provides a unified framework to predict shape descriptors with measures of both forms of uncertainty directly from 3D images. It maintains the end-to-end nature of DeepSSM while providing an accuracy improvement and uncertainty quantification. By predicting and quantifying uncertainty on PCA scores, Uncertain-DeepSSM enables population-level statistical analysis with aleatoric and epistemic uncertainty measures that can be evaluated in a visually interpretable way. In the future, a layer that maps the PCA scores to the set of correspondence points could be added, enabling fine-tuning the network and potentially providing an accuracy improvement over deterministically mapping PCA scores. Uncertain-DeepSSM bypasses the time-intensive and cost-prohibitive steps of traditional SSM while providing the safety measures necessary to use deep network predictions in clinical settings. Thus, this advancement has the potential to improve medical standards and increase patient accessibility.

Acknowledgement. This work was supported by the National Institutes of Health under grant numbers NIBIB-U24EB029011, NIAMS-R01AR076120, NHLBI-R01HL135568, and NIGMS-P41GM103545. The content is solely the responsibility of the authors and does not necessarily represent the official views of the National Institutes of Health. MRI scans and segmentation were obtained retrospectively from the AFib database at the University of Utah. The authors would like to thank the Division of Cardiovascular Medicine (data were collected under Nassir Marrouche, MD, oversight and currently managed by Brent Wilson, MD, PhD) at the University of Utah for providing the left atrium MRI scans and their corresponding segmentations.

References

1. Baccetti, T., Franchi, L., McNamara, J.: Thin-plate spline analysis of treatment effects of rapid maxillary expansion and face mask therapy in early class III malocclusions. Eur. J. Orthod. **21**(3), 275–281 (1999)
2. Bhalodia, R., Dvoracek, L.A., Ayyash, A.M., Kavan, L., Whitaker, R., Goldstein, J.A.: Quantifying the severity of metopic craniosynostosis: a pilot study application of machine learning in craniofacial surgery. J. Craniofac. Surg. **31**, 697–701 (2020)
3. Bhalodia, R., Elhabian, S.Y., Kavan, L., Whitaker, R.T.: Deepssm: a deep learning framework for statistical shape modeling from raw images. CoRR abs/1810.00111 (2018). http://arxiv.org/abs/1810.00111
4. Bhalodia, R., et al.: Deep learning for end-to-end atrial fibrillation recurrence estimation. In: Computing in Cardiology, CinC 2018, Maastricht, The Netherlands, 23–26 September 2018 (2018)
5. Bookstein, F.L.: Principal warps: thin-plate splines and the decomposition of deformations. IEEE Trans. Pattern Anal. Mach. Intell. **11**(6), 567–585 (1989)
6. Bryan, R., Nair, P.B., Taylor, M.: Use of a statistical model of the whole femur in a large scale, multi-model study of femoral neck fracture risk. J. Biomech. **42**(13), 2171–2176 (2009)

7. Cates, J., et al.: Computational shape models characterize shape change of the left atrium in atrial fibrillation. Clin. Med. Insights Cardiol. **8**, CMC-S15710 (2014)
8. Cates, J., Elhabian, S., Whitaker, R.: Shapeworks: particle-based shape correspondence and visualization software. In: Statistical Shape and Deformation Analysis, pp. 257–298. Elsevier (2017)
9. Cates, J., Fletcher, P.T., Styner, M., Shenton, M., Whitaker, R.: Shape modeling and analysis with entropy-based particle systems. In: Karssemeijer, N., Lelieveldt, B. (eds.) IPMI 2007. LNCS, vol. 4584, pp. 333–345. Springer, Heidelberg (2007). https://doi.org/10.1007/978-3-540-73273-0_28
10. Chang, J., Fisher, J.W.: Efficient MCMC sampling with implicit shape representations. In: CVPR 2011, pp. 2081–2088. IEEE (2011)
11. Davies, R.H., Twining, C.J., Cootes, T.F., Waterton, J.C., Taylor, C.J.: A minimum description length approach to statistical shape modeling. IEEE Trans. Med. Imaging **21**(5), 525–537 (2002). https://doi.org/10.1109/TMI.2002.1009388
12. Denker, J.S., LeCun, Y.: Transforming neural-net output levels to probability distributions. In: Advances in Neural Information Processing Systems, pp. 853–859 (1991)
13. Durrleman, S., et al.: Morphometry of anatomical shape complexes with dense deformations and sparse parameters. NeuroImage **101**, 35–49 (2014)
14. Gal, Y.: Uncertainty in deep learning. Univ. Camb. **1**, 3 (2016)
15. Gal, Y., Ghahramani, Z.: Bayesian convolutional neural networks with bernoulli approximate variational inference. arXiv preprint arXiv:1506.02158 (2015)
16. Gal, Y., Ghahramani, Z.: Dropout as a Bayesian approximation: representing model uncertainty in deep learning. In: International Conference on Machine Learning, pp. 1050–1059 (2016)
17. Galloway, F., et al.: A large scale finite element study of a cementless osseointegrated tibial tray. J. Biomech. **46**(11), 1900–1906 (2013)
18. Gardner, G., Morris, A., Higuchi, K., MacLeod, R., Cates, J.: A point-correspondence approach to describing the distribution of image features on anatomical surfaces, with application to atrial fibrillation. In: 2013 IEEE 10th International Symposium on Biomedical Imaging, pp. 226–229, April 2013. https://doi.org/10.1109/ISBI.2013.6556453
19. Gerig, G., Styner, M., Jones, D., Weinberger, D., Lieberman, J.: Shape analysis of brain ventricles using spharm. In: Proceedings IEEE Workshop on Mathematical Methods in Biomedical Image Analysis (MMBIA 2001), pp. 171–178 (2001). https://doi.org/10.1109/MMBIA.2001.991731
20. Gielis, J.: A generic transformation that unifies a wide range of natural and abstract shapes (2013)
21. Glorot, X., Bengio, Y.: Understanding the difficulty of training deep feedforward neural networks. In: Proceedings of the Thirteenth International Conference on Artificial Intelligence and Statistics. Proceedings of Machine Learning Research, vol. 9, pp. 249–256. PMLR, 13–15 May 2010
22. Harris, M.D., Datar, M., Whitaker, R.T., Jurrus, E.R., Peters, C.L., Anderson, A.E.: Statistical shape modeling of cam femoroacetabular impingement. J. Orthop. Res. **31**(10), 1620–1626 (2013). https://doi.org/10.1002/jor.22389
23. He, K., Zhang, X., Ren, S., Sun, J.: Delving deep into rectifiers: surpassing human-level performance on imagenet classification. CoRR abs/1502.01852 (2015). http://arxiv.org/abs/1502.01852
24. Ho, S.Y., Cabrera, J.A., Sanchez-Quintana, D.: Left atrial anatomy revisited. Circ. Arrhythmia Electrophysiol. **5**(1), 220–228 (2012)

25. Huang, W., Bridge, C.P., Noble, J.A., Zisserman, A.: Temporal heartnet: towards human-level automatic analysis of fetal cardiac screening video. In: Descoteaux, M., Maier-Hein, L., Franz, A., Jannin, P., Collins, D., Duchesne, S. (eds.) MICCAI 2017. LNCS, vol. 10434, pp. 341–349. Springer International Publishing, Cham (2017). https://doi.org/10.1007/978-3-319-66185-8_39
26. Joshi, S.C., Miller, M.I.: Landmark matching via large deformation diffeomorphisms. IEEE Trans. Image Process. **9**(8), 1357–1370 (2000)
27. Joshi, S., Davis, B., Jomier, M., Gerig, G.: Unbiased diffeomorphic atlas construction for computational anatomy. NeuroImage **23**(Supplement1), S151–S160 (2004)
28. Kendall, A., Gal, Y.: What uncertainties do we need in Bayesian deep learning for computer vision? CoRR abs/1703.04977 (2017). http://arxiv.org/abs/1703.04977
29. Kingma, D., Ba, J.: Adam: a method for stochastic optimization. In: International Conference on Learning Representations (2014)
30. Kiureghian, A.D., Ditlevsen, O.D.: Aleatory or epistemic? Does it matter? (2009)
31. Kozic, N., et al.: Optimisation of orthopaedic implant design using statistical shape space analysis based on level sets. Med. Image Anal. **14**(3), 265–275 (2010)
32. Kwon, Y., Won, J.H., Kim, B.J., Paik, M.C.: Uncertainty quantification using Bayesian neural networks in classification: application to ischemic stroke lesion segmentation (2018)
33. Lamecker, H., Lange, T., Seebass, M.: A statistical shape model for the liver. In: Dohi, T., Kikinis, R. (eds.) MICCAI 2002. LNCS, vol. 2489, pp. 421–427. Springer, Heidelberg (2002). https://doi.org/10.1007/3-540-45787-9_53
34. Lê, M., Unkelbach, J., Ayache, N., Delingette, H.: Sampling image segmentations for uncertainty quantification. Med. Image Anal. **34**, 42–51 (2016)
35. Le, Q.V., Smola, A.J., Canu, S.: Heteroscedastic Gaussian process regression. In: Proceedings of the 22nd International Conference on Machine Learning, pp. 489–496 (2005)
36. Li, X., Chen, S., Hu, X., Yang, J.: Understanding the disharmony between dropout and batch normalization by variance shift. CoRR abs/1801.05134 (2018). http://arxiv.org/abs/1801.05134
37. MacKay, D.J.: A practical Bayesian framework for backpropagation networks. Neural Comput. **4**(3), 448–472 (1992)
38. Milletari, F., Rothberg, A., Jia, J., Sofka, M.: Integrating statistical prior knowledge into convolutional neural networks. In: Descoteaux, M., Maier-Hein, L., Franz, A., Jannin, P., Collins, D.L., Duchesne, S. (eds.) MICCAI 2017. LNCS, vol. 10433, pp. 161–168. Springer International Publishing, Cham (2017). https://doi.org/10.1007/978-3-319-66182-7_19
39. Moghaddam, B., Pentland, A.: Probabilistic visual learning for object representation. IEEE Trans. Pattern Anal. Mach. Intell. **19**(7), 696–710 (1997)
40. Nix, D.A., Weigend, A.S.: Estimating the mean and variance of the target probability distribution. In: Proceedings of 1994 IEEE International Conference on Neural Networks (ICNN 1994), vol. 1, pp. 55–60. IEEE (1994)
41. Oktay, O., et al.: Anatomically constrained neural networks (ACNN): application to cardiac image enhancement and segmentation. CoRR abs/1705.08302 (2017). http://arxiv.org/abs/1705.08302
42. Reinhold, J.C., et al.: Validating uncertainty in medical image translation. In: 2020 IEEE 17th International Symposium on Biomedical Imaging (ISBI), pp. 95–98. IEEE (2020)
43. Sarkalkan, N., Weinans, H., Zadpoor, A.A.: Statistical shape and appearance models of bones. Bone **60**, 129–140 (2014)

44. Selvan, R., Faye, F., Middleton, J., Pai, A.: Uncertainty quantification in medical image segmentation with normalizing flows. arXiv preprint arXiv:2006.02683 (2020)
45. Styner, M., et al.: Framework for the statistical shape analysis of brain structures using SPHARM-PDM (2006)
46. Tóthová, K., et al.: Uncertainty quantification in cnn-based surface prediction using shape priors. CoRR abs/1807.11272 (2018). http://arxiv.org/abs/1807.11272
47. Wang, D., Shi, L., Griffith, J.F., Qin, L., Yew, D.T., Riggs, C.M.: Comprehensive surface-based morphometry reveals the association of fracture risk and bone geometry. J. Orthop. Res. **30**(8), 1277–1284 (2012)
48. Xie, J., Dai, G., Zhu, F., Wong, E.K., Fang, Y.: Deepshape: deep-learned shape descriptor for 3d shape retrieval. IEEE Trans. Pattern Anal. Mach. Intell. **39**(7), 1335–1345 (2017)
49. Zhao, Z., et al.: Hippocampus shape analysis and late-life depression. PLoS One **3**(3), e1837 (2008)
50. Zheng, Y., Liu, D., Georgescu, B., Nguyen, H., Comaniciu, D.: 3D deep learning for efficient and robust landmark detection in volumetric data. In: Navab, N., Hornegger, J., Wells, W.M., Frangi, A.F. (eds.) MICCAI 2015. LNCS, vol. 9349, pp. 565–572. Springer, Cham (2015). https://doi.org/10.1007/978-3-319-24553-9_69

D-net: Siamese Based Network for Arbitrarily Oriented Volume Alignment

Jian-Qing Zheng[1]([✉])(ID), Ngee Han Lim[1](ID), and Bartłomiej W. Papież[2](ID)

[1] Kennedy Institute of Rheumatology, Nuffield Department of Rheumatology, Orthapaedics and Musculoskeletal Sciences, University of Oxford, Oxford, UK
{jianqing.zheng,han.lim}@kennedy.ox.ac.uk
[2] Big Data Institute, Li Ka Shing Centre for Health Information and Discovery, University of Oxford, Oxford, UK
bartlomiej.papiez@bdi.ox.ac.uk

Abstract. Alignment of contrast and non contrast-enhanced imaging is essential for quantification of changes in several biomedical applications. In particular, the extraction of cartilage shape from contrast-enhanced Computed Tomography (CT) of tibiae requires accurate alignment of the bone, currently performed manually. Existing deep learning-based methods for alignment require a common template or are limited in rotation range. Therefore, we present a novel network, D-net, to estimate arbitrary rotation and translation between 3D CT scans that additionally does not require a prior template. D-net is an extension to the branched Siamese encoder-decoder structure connected by new mutual, non-local links, which efficiently capture long-range connections of similar features between two branches. The 3D supervised network is trained and validated using preclinical CT scans of mouse tibiae with and without contrast enhancement in cartilage. The presented results show a significant improvement in the estimation of CT alignment, outperforming the current comparable methods.

Keywords: Image registration · Deep learning · Mutual non-local link · Siamese encoder decoder

1 Introduction

It is currently impossible to accurately quantify the damage to cartilage during the progression of disease in small animal models of osteoarthritis. Visualisation of cartilage in Computed Tomography (CT) requires a contrast agent. The preclinical development of such a contrast agent [9] has highlighted the problem of accurate cartilage shape extraction from contrasted images; the partial volume effect and adjacency to bone necessitates the use of pre-and post-contrasted CT images. In preclinical scanners, the animal can be placed in various unsystematic (i.e. arbitrary) positions during the acquisition. Tibial cartilage shape may be extracted from the contrast enhanced image by subtracting the non-contrasted

© Springer Nature Switzerland AG 2020
M. Reuter et al. (Eds.): ShapeMI 2020, LNCS 12474, pp. 73–84, 2020.
https://doi.org/10.1007/978-3-030-61056-2_6

scan but this requires accurate alignment of the tibial bone. However, current semi-accurate manual alignment using *ImageJ* requires over 1 h and is prone to error, calling for an automated and accurate method to estimate rigid transformation between 3D volumes acquired in the preclinical setup.

The standardised protocols for image acquisition in clinical scanners means that the range of rotation and translation required to register scans are small and the bigger challenge is to perform deformable registration, especially in between image modalities [14]. In pre-clinical studies, protocols are usually study and machine specific. The limbs of mice are particularly challenging as they may be extended or tucked, dependent on posture (prone, supine and on the side). Post-mortem *ex vivo* tissue may also be scanned in fixative solution, which increases the variability of orientations and positions of scans. Our initial dataset is comprised of such *ex vivo* tissue which will later be used to validate the *in vivo* scans. Thus estimation of large-range rigid transformations is required.

Classic approaches to preclinical image alignment [1] used state-of-the-art, iterative image registration with a similarity measure capturing intensity changes caused by contrast and an appropriate transformation model. However, such approaches are easily trapped in local minimum especially when large translation or rotation is present. More recently, deep learning approaches [8,10,15] were employed to improve the performance of iterative image registration algorithms, however, the slow performance and dependency on initialization motivates one-step transformation estimation via regression [4,5]. For example, two-branch Siamese Encoder (SE) used to learn similarity measure between two images, was applied to 2D brain images alignment [16]. A convolution neural network called AIRNet [2] was used for affine registration of 3D brain Magnetic Resonance Imaging (MRI) with dense convolution layers as SE. The SE structure was also used within the framework of deformable image registration in [17,18] to estimate an initial, affine transformation between two volumes. Alternatively, affine transformation can be estimated using the Global-net [6] with the input images being concatenated and fed into an one-branch encoder. Despite the success of the previous approaches, the capture range of rotation is heavily limited between ±15° [16] and ±45.84° (0.8 rad) [2], yielding unsatisfactory results in preclinical imaging acquisition setup (shown in Sect. 3). The 3D pose estimation of arbitrary oriented subject was presented in [13] but it requires a prior standard template, which is not available for preclinical cartilage imaging.

In this paper, a new architecture, D-net, is proposed for estimation of arbitrary rigid transformation based on a Siamese Encoder Decoder (SED) with novel Mutual Non-local Links (MNL) between two Siamese branches, as described in Sect. 2.1 and Sect. 2.2. Data collection and experiment design are described in Sect. 2.3 and Sect. 2.4 respectively. Experimental results are shown in Sect. 3, discussed and concluded in Sect. 4.

The contributions of this work are as follows. We propose a new network with SED used for first time for rigid registration and we present a concept of MNL showing significantly improved performance on 3D volume alignment. Our network achieves consistent accuracy for wide range of volume orientations apparent in challenging preclinical data set while it does not require prior atlas or template.

2 Methodology

The objective of 3D image registration is to estimate the transformation $f : \mathbb{R}^s \to \mathbb{R}^s$, $\boldsymbol{X}^f \mapsto \boldsymbol{X}^m$ between a fixed volume $\boldsymbol{X}^f \in \mathbb{R}^s$ and a moving volume $\boldsymbol{X}^m \in \mathbb{R}^s$, where $s = d \times h \times w$, and d, h, w are the thickness, height, and width. For 3D rigid registration, the transformation $f_\theta := [\boldsymbol{R}, \boldsymbol{t}] \in SE(3)$, consists of rotation $\boldsymbol{R} \in SO(3)$ and translation $\boldsymbol{t} \in \mathbb{R}^3$, with the parameters $\theta = [\theta^r, \theta^t] \in \mathbb{R}^{12}$ including θ^t and θ^r for translation and rotation. The task of the networks in registration is to estimate θ from the two preprocessed volumes $\tilde{\boldsymbol{X}}^f$ and $\tilde{\boldsymbol{X}}^m$ by networks' mapping $g : (\tilde{\boldsymbol{X}}^f, \tilde{\boldsymbol{X}}^m) \mapsto \hat{\theta}$, where $\hat{\theta} \in \mathbb{R}^{12}$ are the parameters estimated by networks.

As $\theta^r \in \mathbb{R}^9$ is redundant for rotation, the 3D orthogonalization mapping of 6D rotation representation [20] is used as $O : \mathbb{R}^6 \to SO(3)$, $\theta^r_{1-6} \mapsto \boldsymbol{R}$, calculated by:

$$O(\theta^r_{1-6}) = [\boldsymbol{r}_1 \ \boldsymbol{r}_2 \ \boldsymbol{r}_3] := \begin{bmatrix} N([\theta^r_{1-3}]^\top) \\ N([\theta^r_{4-6}]^\top - (\boldsymbol{r}_1^\top \cdot [\theta^r_{4-6}])\boldsymbol{r}_1^\top) \\ \det([\boldsymbol{r}_1 \ \boldsymbol{r}_2 \ \boldsymbol{e}]) \end{bmatrix}^\top \tag{1}$$

where $[\theta_{i-j}] \in \mathbb{R}^{j-i}$ denotes a column vector consist of θ_{i-j}, $N(\cdot)$ denotes a Euclidean normalization function, $\det(\cdot)$ denotes a determinant calculation, $\boldsymbol{e}$ is a vector of the 3 canonical basis vectors of the 3D Euclidean space. This mapping keeps the continuous representation of 3D rotation and is equivalent to Gram-Schmidt process for a rotation in right handed coordinate system but just requires 6 input values.

Thus the rigid transformation can be estimated by: $\hat{f}_{\hat{\theta}} = [O(\hat{\theta}^r), \hat{\theta}^t]$.

2.1 D-net Architecture

D-net consists of SE part, decoder part, and regression part, and its schematic architecture is shown in Fig. 1(a). Similar structure of SED was applied to segmentation [7] and tracking [3], but with different connection structure between contracting and expansive parts comparing to D-net. The SE in the D-net includes two branches of six Residual-down-sampling (Res-down) blocks with shared parameters. Four pairs of the Res-down blocks are linked by MNL and detailed in the Fig. 1(b). In MNL, two matching matrices, from left branch to right and the inverse, are computed by dot product of each pair of voxels' feature vectors. The matrices from two branches are normalized via softmax to correspond and connect the voxels of the feature maps between two branches. MNL, therefore, captures the long-range connection of similar high and low level features between two branches.

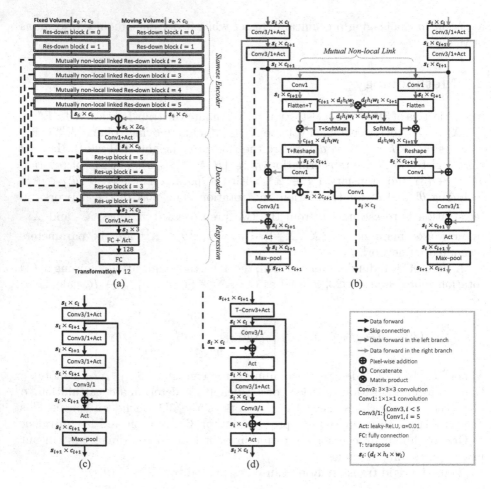

Fig. 1. The architecture of (a) D-net, (b) the novel Mutually non-local linked Res-down block, (c) Res-down block, and (d) Res-up block.

The details of the Res-down blocks are illustrated in the Fig. 1(c). The decoder part of D-net includes four Residual-up-sampling (Res-up) blocks receiving skip connections from the corresponding Mutual Non-local Linked Res-down blocks, shown in the Fig. 1(d). The regression part of D-net includes two fully connected layers with 128 and 12 neurons for 12 transformation parameters. In Fig. 1, i is the block number, $\boldsymbol{d} = (d_0 \cdots d_6) \in \mathbb{Z}_+^7$, $\boldsymbol{h} = (h_0 \cdots h_6) \in \mathbb{Z}_+^7$, $\boldsymbol{w} = (w_0 \cdots w_6) \in \mathbb{Z}_+^7$ and $\boldsymbol{c} = (c_0 \cdots c_6) \in \mathbb{Z}_+^7$ denote the sequences of thickness, heights, widths, and channel number of input volume and feature maps for each branch respectively.

2.2   Mutual Non-local Link

An approach with non-local links was presented in a classic image registration, where non-local motion was estimated using a range of spatial scales, naturally captured by graph representation [12]. Similarly, a concept of unique matching between a pair of voxels by weighting function and mutual saliency was previously shown in [11]. Here, the deep-learning design of MNL was inspired by the Self Non-local Link (SNL) on one branch proposed in [19]. In this section we provide the general definition of MNL:

$$\begin{cases} y_k^{\mathrm{m2f}} := \dfrac{\sum_{\forall j} \phi(x_k^{\mathrm{f}}, x_j^{\mathrm{m}})\psi(x_j^{\mathrm{m}})}{\sum_{\forall j} \phi(x_k^{\mathrm{f}}, x_j^{\mathrm{m}})} \\ y_k^{\mathrm{f2m}} := \dfrac{\sum_{\forall j} \phi(x_k^{\mathrm{m}}, x_j^{\mathrm{f}})\psi(x_j^{\mathrm{f}})}{\sum_{\forall j} \phi(x_k^{\mathrm{f}}, x_j^{\mathrm{m}})} \end{cases} \tag{2}$$

where j, k are the indices of the position in a feature map, $x^{\mathrm{f}}, x^{\mathrm{m}}$ are the input signals from two branches, $y^{\mathrm{m2f}}, y^{\mathrm{f2m}}$ are the output signals from this block, ϕ is the similarity measurement function, ψ is a unary function.

The instantiated MNL in D-net is based on embedded Gaussian similarity representation with

$$\phi(x_1, x_2) := e^{y_1^\top W^\top W x_2} \tag{3}$$

and

$$\psi(x) := W x, \tag{4}$$

where W is a matrix of trainable weights.

2.3   Data Collection and Pre-processing

A total of 100 ex-vivo micro CT scans of tibiae from 50 mice were acquired using *Perkin Elmer*, Quantum FX with a resolution of $10 \times 10 \times 10\,\mu\mathrm{m}^3$/vox, volume size of $512 \times 512 \times 512$. Scans varied from 0 week to 20 weeks post osteoarthritic surgery. Each tibiae was scanned once pre-contrast and once post-contrast with washout. Due to the handling, the muscles, solution and small broken bone fragments are displaced differently from the tibial bone.

To remove the possible influence of muscle and solution on alignment of tibial bones, we preprocessed the collected data by thresholding and normalizing with mapping $T : \mathbb{R}^s \to \mathbb{R}^s, X \mapsto \tilde{X}$ by:

$$\tilde{x}_{ijk} = \frac{\mathrm{ReLU}(x_{ijk} - x_{\mathrm{th}})}{x_{\max} - x_{\mathrm{th}}}, \tag{5}$$

where $\tilde{x}_{ijk}$ is the entry of $\tilde{X}$, $x_{\max}$ is the maximum intensity in the dataset and x_{th} is the threshold value set as 2000 because of the high density of background solution. Finally, because of data size, the input volumes are sub-sampled with linear interpolation to $80 \times 80 \times 80\,\mu\mathrm{m}^3$/vox. The CT slices of two exemplar subjects are shown in Fig. 2.

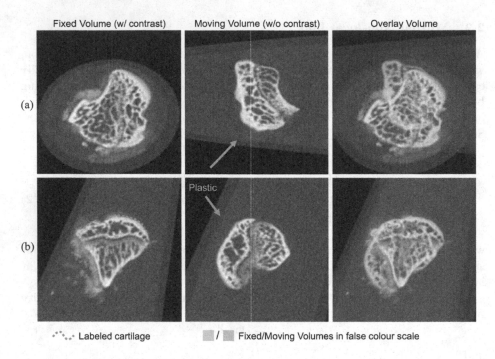

Fig. 2. The tibial CT slices (at original resolution) of two exemplar subjects show the large ranges of spatial transformation between fixed and moving volumes.

In the training dataset, each CT volume is transformed to synthesize the fixed and the moving volumes with uniformly distributed random translation $\sim \mathcal{U}(-0.64, 0.64)$mm in 3D, varying by $\frac{1}{4}$ of the whole volume size, and random rotation with angle $\sim \mathcal{U}(-\pi, \pi)$ around random axes uniformly distributed in 3D sphere surface. To enlarge the training dataset, data augmentation including intensity scale transforming and Gaussian noise is applied, where the intensity scale coefficient is $\sim \mathcal{U}(0.95, 1.05)$ and each voxel is added with random variable $\sim \mathcal{N}(0, 0.001)$.

2.4   Training and Validation

The loss function in terms of θ and $\hat{\theta}$ is calculated as:

$$\mathcal{L} = \alpha \frac{\|\theta^{\mathrm{t}} - \hat{\theta}^{\mathrm{t}}\|_2^2}{\|\theta^{\mathrm{t}}\|_2^2 + \epsilon} + \beta\|\theta^{\mathrm{r}} - \hat{\theta}^{\mathrm{r}}\|_2^2 \tag{6}$$

where $\| \cdot \|_2 := \sqrt{(\sum \cdot)^2}$ is Euclidean norm, the both weights of relative translation error α and rotation error β are set as $\alpha = \beta = 0.5$, and $\epsilon = 0.01$ to avoid singularity. Momentum Stochastic Gradient Descend was applied with the learning rate 0.0001 and the learning momentum 0.9.

We split our CT data set into two folders (A and B) each containing pairs of contrast and non contrast-enhanced CTs from the same mouse, and two-fold cross validation was performed on different mouse. Furthermore, we performed four validation strategies: (S1) training on all data from folder A, and testing on B; (S2) training on all data from B, testing on A; (S3) training on non contrast-enhanced data from A, and testing on all data from B; (S4) training on non contrast-enhanced data from B, and testing on all data from A. Training on non contrast-enhanced data was performed to check whether our network can be used for alignment of follow-up contrast-enhanced data, when only the baseline data are available for training, thus modeling real scenario of data acquisition. For the validation strategy, known transformation (as described in Sect. 2.3) was applied to volume from the folder to create a pair of CTs in synthetic test, and in real test, each pair of corresponding contrasted and non-contrasted CT was registered. The synthetic test includes rotation test with fixed translation $(0.4\ 0.4\ 0.4)$ mm and 11 angles rotation uniformly ranges from $-\pi$ to π around axis $(\frac{1}{\sqrt{3}}\ \frac{1}{\sqrt{3}}\ \frac{1}{\sqrt{3}})$ as well as translation test with fixed $\frac{\pi}{2}$ rotation around axis $(\frac{1}{\sqrt{3}}\ \frac{1}{\sqrt{3}}\ \frac{1}{\sqrt{3}})$ and 11 translation uniformly ranges from $-\frac{\sqrt{3}}{2}$ mm to $\frac{\sqrt{3}}{2}$ mm along the axis $(\frac{1}{\sqrt{3}}\ \frac{1}{\sqrt{3}}\ \frac{1}{\sqrt{3}})$.

2.5 Comparison and Evaluation

D-net was compared with other relevant image registration approaches:

- SITK: Simple ITK with metric Joint Histogram Mutual Information and optimizer Regular Step Gradient Descent with gradient tolerance 0.0001, max iter number 10k and learning rate 1.
- ME (Mixed Encoder): The "Global-net" [6] concatenating the two input volumes together and feeding into one mixed branch. All the architecture settings are set with default values, with $d = h = w = (64, 32, 16, 8, 4)$ and $c = (1, 16, 32, 64, 128)$.
- SE (Siamese Encoder): An architecture employing two branches of 6 Res-down blocks for SE and two fully connected layers for regression, a similar structure was used in [17] but without residual structure and fewer down-sampling blocks.
- SED (Siamese Encoder Decoder): A proposed architecture inserting 4 Res-up blocks into SE between the SE and regression parts, with skip connection from the 4 latter Res-down block of SE.
- SNL-SED (Self Non-local Linked - Siamese Encoder Decoder): A proposed SED architecture with the 4 latter Res-down blocks self non-locally linked by the Embedded Gaussian similarity based - non-local block [19] in each branch.

SITK, ME and SE are previously published methods; while SED and SNL-SED are transitional forms towards D-Net and their impact is separately validated; SE, SED SNL-SED and D-net are validated with $d = h = w = (64, 32, 16, 8, 4, 2, 1)$ and $c = (1, 16, 32, 64, 64, 64, 64)$.

Criteria. The Euclidean distance of Translation Error (TE) between the predicted and expected translation, $TE := \|\theta^t - \hat{\theta}^t\|_2$, and the Rotation Error (RE) between the predicted and expected rotation, $RE = \arccos(\frac{\operatorname{tr}(R^\top O(\hat{\theta}^r))-1}{2})$, are calculated for synthetic tests. Since the ground truth for real examples are unknown, the Dice Similarity Coefficient (DSC) between the cortical bone segmented from contrasted and non-contrasted tibial CT is calculated for both synthetic and real tests.

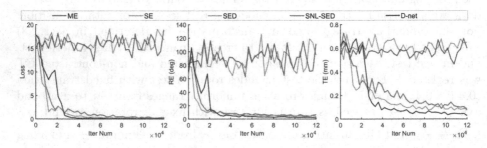

Fig. 3. The training curves exemplified with (S1) left: Loss values, middle: Rotation errors (RE) and right: Translation Errors (TE) shows SED, SNL-SED and D-net are trainable across range of translations and rotations.

3   Results

All networks were trained for 120k iterations. In all training strategies used, ME and SE failed to converge, whereas SED, SNL-SED and D-net were trainable (exemplified in Fig. 3).

The results of rotation test and translation test with validation strategy (S1&S2) are shown in Fig. 4(a) and Fig. 4(b), where only D-net achieves the sub-voxel average TE in the rotation and translation tests. The performance of ME and especially SE is sensitive to the initial translation and rotation as shown in Fig. 4(a)-middle and Fig. 4(b)-left because they intend to predict a small range transformations for any input volumes. DSCs for 30 subjects in real test with strategy (S1&S2) are shown in Fig. 4(c), where the SED increases DSC by 0.1−0.5 from ME and D-net further raises DSC by 0.1−0.4 from SED. The average DSC for strategy (S1)−(S4) in rotation test and real test are plotted in Fig. 4(d). It shows average DSC of D-net is higher than all others in both rotation and real test but is slightly lower in real test compared with rotation test.

The tibial bone shapes of two registration examples for ME, SED, and D-net from real test are shown in Fig. 5 with the same subjects previously shown in Fig. 2. The figure illustrates bone fragments and segmentation difference caused by the varying intensity influenced by contrast, decreasing the DSC values and making registration of preclinical tibia data particularly challenging. Visual

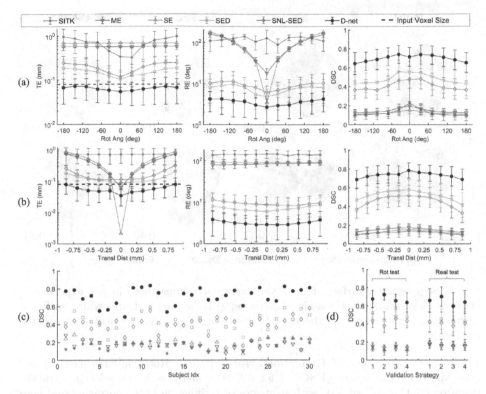

Fig. 4. D-net outperforms all other methods exemplified by (S1) in (a) rotation (rot) test, (b) translation test, and (c) real test exemplified by 30 subjects, with Translation Errors (TE), Rotation Errors (RE) and DSC, avg±std; (d) DSC, avg±std in rot test and real test with validation strategies (S1)−(S4).

results in Fig. 5 confirms that D-Net performs robustly and this is further supported by the quantitative results shown in Table 1, where TE, RE, and DSC for all methods are presented.

Comparing with others, D-net achieves the lowest TE and RE and highest DSC with consistent performance across range of rotations. Using two-way Analysis of Variance (ANOVA) in rotation and translation tests and one-way ANOVA in real test, D-net significantly outperforms all other approaches with $p < 10^{-4}$ on TE, RE and DSC in rotation and translation tests and on DSC in real test by strategy (S1&S2) and (S3&S4); SED significantly outperforms SE, ME and SITK with $p < 10^{-4}$ on TE, RE and DSC in rotation and translation test and on DSC in real test by all the strategies.

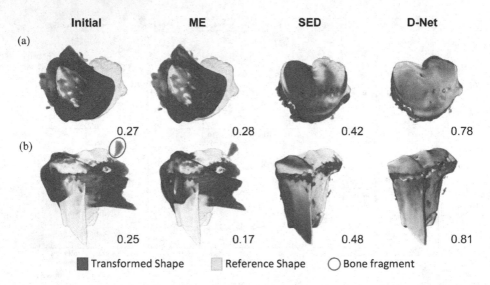

Fig. 5. Segmentation surfaces for the two exemplar volumes used in real experiment. D-net achieve the most plausible registration (overlapping red and white surfaces) with highest DSC shown at the right bottom corner (images are shown at original resolution).

Table 1. Average values of Translation Error (TE/μm), Rotation Error (RE/°) and Dice Similarity Coefficient (DSC/%) for different methods in Translation (Transl), Rotation (Rot) and Real test, with strategies of training on both contrasted and non-contrasted data (S1&S2) and just non-contrasted data (S3&S4)

		SITK	ME		SE		SED		SNL-SED		D-net	
Variable No.	-		0.7M		9.2M		5.2M		4.9M		4.9M	
Strategy(S)	-		1 & 2	3 & 4	1 & 2	3 & 4	1 & 2	3 & 4	1 & 2	3 & 4	1 & 2	3 & 4
Transl	TE	710.5	406.4	395.1	472.6	472.5	124.3	132.7	163.4	146.2	**55.8**	**77.7**
	RE	135.9	82.7	83.4	89.8	89.2	7.1	8.0	9.3	9.7	**3.2**	**4.2**
	DSC	12.98	15.04	14.94	12.86	13.42	52.28	51.08	45.61	47.69	**73.57**	**66.44**
Rot	TE	696.2	575.4	569.4	692.4	692.4	154.4	144.1	201.5	175.7	**65.3**	**81.0**
	RE	119.6	92.6	95.1	98.5	98.0	7.4	8.1	9.1	9.5	**3.6**	**4.5**
	DSC	12.75	14.54	14.54	12.22	12.75	48.28	47.97	40.88	44.69	**69.83**	**64.64**
Real	DSC	16.91	16.75	15.16	18.86	17.23	43.66	42.52	41.85	39.26	**67.92**	**61.68**

4 Discussion and Conclusion

In this paper, we proposed a new network, D-net, and a new structure, Mutual Non-local Link (MNL), for estimation of transformation between CT volumes.

The experimental results shows D-net outperforms other methods and achieves state-of-the-art performance for rigid registration of preclinical mouse CT scans with and without contrast. While ME and SE did not converge during training using full range of rotations, we were able to train them using smaller range of rotations (30°), similarly as in [2,4]. This further shows superiority of D-net at consistently estimating full range of rotations.

The average DSCs of D-net in real test are slightly lower than in synthetic tests potentially due to the difference in segmentation for contrast-enhanced volumes. However, D-net is still able to extract the common features of tibial bone and align two volumes plausibly, showing usefulness for shape extraction of cartilage from contrast-enhaced CT of tibiae.

For the rotation representation, the widely used quaternion, Euler angles and Lee algebra were not applied due to the discontinuity of 3D rotation represented in the real Euclidean space with dimension lower than 5D [20].

In future work, a pipeline for cartilage shape extraction will be further validated for morphological analysis and in application for diagnosis and staging of osteoarthritis, and D-net will be generalized to other modalities to explore inter-subject and inter-modality registration.

Acknowledgement. This work was supported by a Kennedy Trust for Rheumatology Research Studentship, the Centre for OA Pathogenesis Versus Arthritits (Versus Arthritis grant 21621). The authors acknowledge Patricia das Neves Borges as the researcher who collected the preclinical CT dataset, as part of the National Centre for Replacement, Refinement and Reduction of Animals in Research (NC3R grant NC/M000141/1). B. W. Papież acknowledges Rutherford Fund at Health Data Research UK

References

1. Baiker, M., Staring, M., Löwik, C.W.G.M., Reiber, J.H.C., Lelieveldt, B.P.F.: Automated registration of whole-body follow-up MicroCT data of mice. In: Fichtinger, G., Martel, A., Peters, T. (eds.) MICCAI 2011. LNCS, vol. 6892, pp. 516–523. Springer, Heidelberg (2011). https://doi.org/10.1007/978-3-642-23629-7_63

2. Chee, E., Wu, Z.: Airnet: self-supervised affine registration for 3d medical images using neural networks. arXiv preprint arXiv:1810.02583 (2018)

3. Dunnhofer, M., et al.: Siam-U-Net: encoder-decoder siamese network for knee cartilage tracking in ultrasound images. Med. Image Anal. **60**, 101631 (2020)

4. Haskins, G., et al.: Learning deep similarity metric for 3D MR-TRUS image registration. Int. J. Comput. Assist. Radiol. Surg. **14**(3), 417–425 (2019)

5. Haskins, G., Kruger, U., Yan, P.: Deep learning in medical image registration: a survey. Mach. Vis. Appl. **31**(1), 8 (2020)

6. Hu, Y., et al.: Label-driven weakly-supervised learning for multimodal deformable image registration. In: 2018 IEEE 15th International Symposium on Biomedical Imaging (ISBI 2018), pp. 1070–1074. IEEE (2018)

7. Kwon, D., et al.: Siamese U-net with healthy template for accurate segmentation of intracranial hemorrhage. In: Shen, D., et al. (eds.) MICCAI 2019. LNCS, vol. 11766, pp. 848–855. Springer, Cham (2019). https://doi.org/10.1007/978-3-030-32248-9_94

8. Liao, R., et al.: An artificial agent for robust image registration. In: Thirty-First AAAI Conference on Artificial Intelligence (2017)

9. Lim, N.H., Fowkes, M.M.: Radiopaque compound containing diiodotyrosine, 5 Jun 2019, EU Patent EP3490614A1
10. Ma, K., et al.: Multimodal image registration with deep context reinforcement learning. In: Descoteaux, M., Maier-Hein, L., Franz, A., Jannin, P., Collins, D.L., Duchesne, S. (eds.) MICCAI 2017. LNCS, vol. 10433, pp. 240–248. Springer, Cham (2017). https://doi.org/10.1007/978-3-319-66182-7_28
11. Ou, Y., Sotiras, A., Paragios, N., Davatzikos, C.: DRAMMS: deformable registration via attribute matching and mutual-saliency weighting. Med. Image Anal. **15**(4), 622–639 (2011)
12. Papież, B.W., Szmul, A., Grau, V., Brady, J.M., Schnabel, J.A.: Non-local graph-based regularization for deformable image registration. In: Müller, H., et al. (eds.) MICCAI 2019. LNCS, vol. 11766, pp. 199–207. Springer, Cham (2016). https://doi.org/10.1007/978-3-030-32248-9_94
13. Salehi, S.S.M., Khan, S., Erdogmus, D., Gholipour, A.: Real-time deep pose estimation with geodesic loss for image-to-template rigid registration. IEEE TMI **38**(2), 470–481 (2018)
14. Schnabel, J.A., Heinrich, M.P., Papież, B.W., Brady, J.M.: Advances and challenges in deformable image registration: from image fusion to complex motion modelling. Med. Image Anal. **33**, 145–148 (2016)
15. Kwon, D., et al.: Siamese U-net with healthy template for accurate segmentation of intracranial hemorrhage. In: Shen, D., et al. (eds.) MICCAI 2019. LNCS, vol. 11766, pp. 848–855. Springer, Cham (2019). https://doi.org/10.1007/978-3-030-32248-9_94
16. Sloan, J.M., Goatman, K.A., Siebert, J.P.: Learning rigid image registration-utilizing convolutional neural networks for medical image registration. In: 11th International Joint Conference on Biomedical Engineering Systems and Technologies, pp. 89–99. SCITEPRESS-Science and Technology Publications (2018)
17. de Vos, B.D., Berendsen, F.F., Viergever, M.A., Sokooti, H., Staring, M., Išgum, I.: A deep learning framework for unsupervised affine and deformable image registration. Med. Image Anal. **52**, 128–143 (2019)
18. Wang, C., Papanastasiou, G., Chartsias, A., Jacenkow, G., Tsaftaris, S.A., Zhang, H.: FIRE: unsupervised bi-directional inter-modality registration using deep networks. arXiv preprint arXiv:1907.05062 (2019)
19. Wang, X., Girshick, R., Gupta, A., He, K.: Non-local neural networks. In: Proceedings of the IEEE Conference on Computer Vision and Pattern Recognition, pp. 7794–7803 (2018)
20. Zhou, Y., Barnes, C., Lu, J., Yang, J., Li, H.: On the continuity of rotation representations in neural networks. In: Proceedings of the IEEE Conference on Computer Vision and Pattern Recognition, pp. 5745–5753 (2019)

A Method for Semantic Knee Bone and Cartilage Segmentation with Deep 3D Shape Fitting Using Data from the Osteoarthritis Initiative

Justus Schock[1,2,3]([✉]), Marcin Kopaczka[1], Benjamin Agthe[1], Jie Huang[1], Paul Kruse[1], Daniel Truhn[1,3], Stefan Conrad[4], Gerald Antoch[2,3], Christiane Kuhl[1], Sven Nebelung[2,3], and Dorit Merhof[1]

[1] Institute of Imaging and Computer Vision, RWTH Aachen University, Aachen, Germany
justus.schock@lfb.rwth-aachen.de
[2] Insititute of Diagnostic and Interventional Radiology, University Hospital Düsseldorf, Düsseldorf, Germany
[3] Institute of Diagnostic and Interventional Radiology, University Hospital Aachen, Aachen, Germany
[4] Institute of Computer Science, Heinrich Heine University Düsseldorf, Düsseldorf, Germany

Abstract. We present a multistage method for deep semantic segmentation of bone structures based on a landmark-based shape regression and subsequent local segmentation of relevant areas. Our solution covers the entire pipeline from 2D-based pre-segmentation, a method for fast deep 3D shape regression and subsequent patch-based 3D semantic segmentation for final segmentation. Since we perform landmark regression using a statistical shape model, our method is able to fit an arbitrary number of landmarks without increase in model complexity. The algorithm is evaluated on the OAI-ZIB dataset, for which we use the binary masks to generate sets of corresponding landmarks and build a deep statistical shape model. By employing our proposed deep shape fitting, our method achieves the performance of existing high-precision approaches in terms of segmentation accuracy while at the same time drastically reducing computational complexity and improving runtime by a large margin.

Keywords: Knee cartilage segmentation · Statistical shape model · Shape regression · Landmark detection

1 Introduction and Previous Work

In recent years, deep learning has drastically improved the performance of semantic segmentation algorithms thanks to the introduction of fully convolutional

---

J. Schock, M. Kopaczka, S. Nebelung and D. Merhof—Both authors contributed equally to this work.

© Springer Nature Switzerland AG 2020
M. Reuter et al. (Eds.): ShapeMI 2020, LNCS 12474, pp. 85–94, 2020.
https://doi.org/10.1007/978-3-030-61056-2_7

networks [7] and - especially in the biomedical field - the U-Net [10] and its many derivatives which are still the baseline and often best performing methods for numerous segmentation tasks. Given sufficient data and computational power and memory, even basic U-Net architectures yield excellent performance on a large number of segmentation tasks. However, a problem currently faced by many researchers working with volumetric radiological data is the problem that full resolution CT or MRI volumes either do not fit into GPU memory or that the number of manual ground truth annotations is not sufficient for training the network. The latter problem can be addressed by obtaining more annotations with either increased manual work or different augmentation techniques starting with traditional augmentation or wherever possible deep augmentation methods such as (variational) autoencoders [6] or generative adversarial networks [4]. But even with sufficient training data, the model still needs to fit into GPU memory which is currently a limiting factor and often addressed by splitting the input into smaller subvolumes and segmenting these individually. This comes at the cost of context loss. At the same time, often specific parts such as the organ boundary require most attention while segmenting irrelevant volume sections is not required and might increase segmentation time and potentially also lower accuracy. In knee segmentation for osteoarthritis analysis for example, the bone and especially cartilage areas are the most relevant parts. At the same time, the cartilage accounts only for a fraction of the voxels in the volumes.

To focus attention of 3D segmentation networks on relevant areas, several authors have proposed refinement approaches in which a coarse segmentation for relevant area detection is performed first and subsequently only relevant areas are segmented in high quality. While a well-performing method, this approach has usually the downside of either requiring multiple networks or even a non-deep learning method such as statistical shape fitting which drastically increases runtime.

We address this issue by presenting a deep regression method for 3D landmark detection. It allows dense landmark localization that ensures anatomical plausibility by its underlying statistical shape model. At the same time, using a statistical shape model within the network allows transforming the landmark detection task to the task of parametrizing the shape model, thereby allowing the use of a lightweight fully convolutional feature extraction architecture and therefore very high computational efficiency.

Inspired by methods for active shape and appearance model fitting such as [9], several other authors have addressed PCA-based landmark localization with CNNs in medical images. In [8], a method for PCA-based deep landmark detection in ultrasound images has been proposed. The method works on 2D image data and applies a regression-based postprocessing to improve landmark detection precision. In 3D, deep PCA-based landmark detection has been analyzed on prostate MRI data by [5]. The authors proposed a deep convolutional network to predict landmark positions, however with no additional processing to improve accuracy. Furthermore, both approaches were relying on networks with fully connected layers, which increase the number of parameters in the networks and thereby require more training data and time to converge. For example, the authors of [5] used extensive data augmentation to increase the number of training samples. In contrast, we focus on designing a lightweight architecture that allows straightforward training using only the images and landmark data.

We have chosen knee MRI analysis as a recent survey [2] has demonstrated that segmenting the whole knee including femur and tibia as well as the corresponding cartilage is a rather hard task. Segmenting only the cartilage is possible with high accuracy as shown by [11]. Considering the femoral and tibial bone as well, [1] yields the best results employing a shape model with time-consuming optimization during inference. Since our work is methodologically similar, we aim at achieving comparable accuracy with increased efficiency by our proposed deep shape fitting.

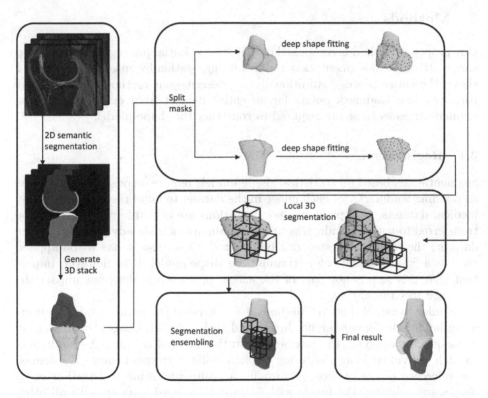

Fig. 1. Schematic of entire algorithm consisting of initial pre-segmentation, landmark generation, local fine segmentations and ensembling of all segmentation results. Note that 3D mask renderings in this figure are used to demonstrate the principle of the method; the actual algorithm for segmentation and shape fitting works entirely on the MRI images.

We make our Python/PyTorch code for deep landmark fitting freely available to allow reproducing our results as well as development of novel landmark detection methods.[1]

2   Materials and Preprocessing

Development and validation of methods were performed on the 507 segmented knee MRI scans from the OAI-ZIB dataset as provided by [1]. Furthermore, data used in the preparation of this article were obtained from the Osteoarthritis Initiative (OAI) database, which is available for public access at http://www. oai.ucsf.edu/.

3   Methods

Our proposed method is comprised of four stages: Initial pre-segmentation, deep shape fitting, local segmentation and result aggregation by ensembling. Figure 1 shows the entire process. Additionally, we describe our method for generating corresponding landmark points for an entire dataset from existing pixel-wise segmentations as these are required to construct the shape model.

3.1   Registration

Segmentations based on statistical shape models require a consistent set of corresponding landmarks for each image in the dataset to build the model. In many medical datasets, only pixel-wise segmentations are available which need to be transferred to a set of landmarks first. If landmarks are already provided by the dataset, the registration step can be omitted. Otherwise it has to be applied once as a preprocessing before training the shape model. It is, however, important that this step is not part of the actual pipeline and does not impact the inference speed in any way.

Similar to established 2D Methods, we generated the point clouds by generating landmarks for a reference image and registering all images to this image. Subsequently, applying the inverse registration allowed warping the landmarks to each individual image, ensuring implicit point correspondences. We defined the reference image by first performing a multiscale affine registration of all masks and selecting the image with highest bone mask overlap with all other images. Since we create separate tibia and femur point sets, the process was performed separately for each of both bones. To account for local deformations that are not covered by affine registration, all transformed masks were additionally registered to the reference with a deformable three-dimensional diffeomorphic daemons registration [12].

---

[1] Code available at https://github.com/justusschock/OAI-Knee-Segmentation.

3.2 Point Cloud and PCA Generation

For point cloud generation, 2000 points were randomly sampled on the reference image's both tibia and femur mask surfaces. Using the displacement data from the deformable registration, the corresponding points on each image were defined by inversely transforming point coordinates from the reference to each individual image. After obtaining consistent point clouds for all images, we build a statistical shape model by first performing procrustes analysis on the landmarks and subsequently applying PCA to the procrustes results. The entire registration procedure is shown in Fig. 2.

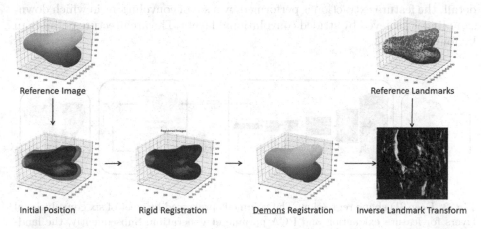

Fig. 2. Landmark generation: All images from the dataset are co-registered to the reference. Subsequently, the landmarks sampled on the reference surface are warped to the respective images by applying the inverse displacement obtained from the registration.

3.3 Shape Model Initialization

To find a suitable initial bone position for shape fitting in each image, we trained a 2D U-net [10] on individual image slices to segment the femur and tibia bones. For training, we found that using the distance-weighted loss presented in [3] yields better accuracy than Dice and cross-entropy losses alone, therefore we trained the U-Net with the sum of cross-entropy and the distance-weighted loss which yielded best results. Subsequently, the segmentation provided by combining the 2D segmentation masks into a 3D segmentation volume was used as initial starting point for the deep shape fitting. To improve GPU memory efficiency and model precision, we cropped the volumes to contain the femur or tibia bone only before forwarding the subvolumes to the shape fitting network.

3.4 Deep 3D Shape Fitting

For deep shape fitting, we propose a CNN architecture that performs regression of shape model parameters. The network consists of a fully convolutional feature extraction stage and a PCA landmark generation stage in which we use the predicted PCA parameters and predicted global transformation parameters to generate landmark locations for all bone landmarks. Fully convolutional architectures have the advantage of allowing building a network with low memory footprint that allows fitting large images into memory. At the same time, the lower number of trainable parameters in comparison with networks that use fully connected layers allows more efficient training and fast regression at runtime. In detail, the feature extraction is performed by a set of convolutions in which downsampling is achieved by strided convolutional layers. The architecture details can be found in Fig. 3.

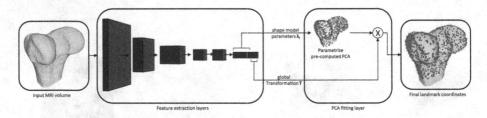

Fig. 3. Deep 3D shape regression. The volume is processed by a set of six convolutional layers for feature extraction and PCA parameter generation. Subsequently, the landmark points are computed in the PCA layer as linear combination of PCA-parametrized shapes and a global landmark transform. The first three layers of the network consist of separated $5 \times 5 \times 5$ convolutions followed by a strided convolution for downsampling and an instance normalization layer. Layers 4 and 5 consist of $3 \times 3 \times 3$ separated convolutions. The final layer in the extraction stage has the same number of channels as the number of PCA parameters and global transform parameters of the shape model, its output is directly mapped to the inputs of the shape layer.

The final convolutional layer of the network uses 1×1-convolutions with the same number of parameters as the PCA components to be predicted together with nine global transformation parameters (three parameters for translation, rotation and scale each). The subsequent proposed PCA layer allows composing a set of landmarks by using the PCA mean, adding a linear combination of its components and subsequently applying the global transformation. Upon network initialization, we load the PCA computed in Sect. 3.2 and select the number of PCA parameters to predict. The output of the network is a list of landmark coordinates in 3D space. This allows efficient network training by using any loss suitable for landmark distance computation such as MSE loss or L1/L2 distance losses. For this work, we used the MSE loss between predicted and ground truth landmark coordinates for training.

3.5   3D Semantic Segmentation and Final Result Computation

After obtaining the 3d landmark positions, we perform a fine segmentation of the cartilage using a local 3D U-Net. A total of 100 3D subvolumes for femur and 30 for tibia are sampled centered at landmark positions in the cartilage regions of the bones (the femur cartilage is much larger, hence the larger number of points) and subsequently merged together to a segmentation of the whole bone structure. The landmarks are selected using distance-based non maximum suppression to ensure equidistant spacing. Finally, result ensembling is performed by adding the class probabilities computed by the 2D and 3D segmentations and subsequently computing the arg max for each voxel, yielding its final class label.

4   Experiments and Results

All experiments were performed on the dataset described in Sect. 2. Of the 507 images, one had to be discarded due to a broken segmentation file and one was discarded as the deformable registration did not converge. From this data, we defined the registration reference as described in Sect. 3.1. To include only images with automatically generated landmarks, the reference image was also discarded from the evaluation. The remaining 504 images were split into a ten-fold cross validation, of which we used eight folds for training, one fold for validation and the remaining fold for testing.

4.1   2D U-Net

We compared our accuracy to the methods presented in [2]. The Dice coefficients of the 2D U-Net (See Table 1) were already outperforming all methods published on OAI data except [1], see Table 1 for details. This shows that for tasks that do not require maximal precision a plain U-Net approach may be sufficient.

4.2   3D Shape Fitting

To assess the landmark detection performance, we trained both individual networks for femur and tibia as well as a variant with two parallel PCA layers to predict both structures simultaneously. The individual networks received subvolumes defined by the 2D segmentation results while the combined network was using the original, full MRI scan. In both cases, the input was resampled to a squared size of $88 \times 88 \times 88$ voxels. An analysis of network performance showed that the achieved distances within the subvolumes were similar for both single- and multi-shape fitting and that the lower multi-shape precision was due to the larger rescaling factor required to resample the shape to the full volume size instead of the size of the individual bone bounding boxes. Since the runtime of a single shape fitting within the whole process is neglectable (the entire shape fitting takes 0.1 s on a Geforce 2080 Ti), we used individual femur and tibia networks to maximize shape fitting precision.

4.3   3D-Unet Refinement

We analyzed the performance of the locally refining 3D U-Nets especially with regard to the number of required volumes. In [1], all landmarks were used as centres for local 3D segmentation. In contrast to this, we focused on cartilage segmentation as this is the most relevant image area and the 2D U-Net already delivered competitive results for bone segmentation. To this end, 100 points for femur cartilage and 30 points for tibia cartilage segmentation were equidistantly selected in the cartilage regions of both bones. Segmentation results are given in Table 1. Quantitative tests with larger numbers of points yielded no improvement, therefore we used the given numbers to maximize computing efficiency.

4.4   Final Results

The results of our algorithm are displayed in Table 1. We were able to achieve and in the case of tibia cartilage even slightly improve the results given in [1]. Thanks to deep shape fitting, our method allows processing of the whole MRI volume on the GPU without any need for CPU computations, making it extremely time-efficient. With a combined overall runtime of all steps of our method of 8 s on a Geforce RTX 2080 Ti, we are able to clearly outperform the runtime of the current reference method (9 min 32 s on a Geforce 980 Ti [1]) with no loss in precision.

Table 1. Dice similarity coefficients for our method, the 2D U-Net and the currently best performing method on this dataset.

Method	Femur	Tibia	Femur cartilage	Tibia cartilage
2D U-Net	98.5	98.5	89.0	84.9
Ours	98.5	98.5	89.9	85.9
Ambellan et al. [1]	98.6	98.5	89.9	85.6

5   Discussion

We were able to demonstrate that our method achieves at least the same accuracy as the current reference at a substantial improvement in computation time. Furthermore, we were able to demonstrate that with a high bone precision, already the 2D U-net trained with a distance-based loss delivers a high segmentation accuracy with low computational and algorithmic requirements. Its high precision coupled with the high processing speed of our method enables using it for fast automated assessment of clinical scans as well as high-throughput analysis of MRI databases. The method has been demonstrated on the OAI knee MRI database, however it can be applied to any 3D segmentation of structures that can be well described by a 3D shape model, allowing widespread use for muscosceletal and organ segmentation tasks.

6 Conclusion

We presented an algorithm for time- and memory-efficient 3D segmentation based on refining 2D segmentations with local 3D segmentations which are positioned in relevant areas of the volume using a deep shape model. Our method works entirely on the GPU and eliminates the need for explicit iterative shape fitting that is currently used in similar approaches. Results show that we achieve state-of-the-art precision while at the same time reducing the computation time from several minutes to a few seconds.

Acknowledgements. We thank The OAI initiative for providing Access to their image data and Felix Ambellan from ZIB Berlin for making the segmentations publically available. The OAI is a public-private partnership comprised of five contracts (N01-AR-2-2258; N01-AR-2-2259; N01-AR-2-2260; N01-AR-2-2261; N01-AR-2-2262) funded by the National Institutes of Health, a branch of the Department of Health and Human Services, and conducted by the OAI Study Investigators. Private funding partners include Merck Research Laboratories; Novartis Pharmaceuticals Corporation, GlaxoSmithKline; and Pfizer, Inc. Private sector funding for the OAI is managed by the Foundation for the National Institutes of Health. This manuscript was prepared using an OAI public use data set and does not necessarily reflect the opinions or views of the OAI investigators, the NIH, or the private funding partners.

References

1. Ambellan, F., Tack, A., Ehlke, M., Zachow, S.: Automated segmentation of knee bone and cartilage combining statistical shape knowledge and convolutional neural networks: data from the osteoarthritis initiative. Med. Image Anal. **52** (2018). https://doi.org/10.1016/j.media.2018.11.009
2. Bonaretti, S., Gold, G.E., Beaupre, G.S.: pyKNEEr: an image analysis workflow for open and reproducible research on femoral knee cartilage. PLoS ONE **15**(1), 1–19 (2020). https://doi.org/10.1371/journal.pone.0226501
3. Eschweiler, D., Klose, T., Müller-Fouarge, F.N., Kopaczka, M., Stegmaier, J.: Towards annotation-free segmentation of fluorescently labeled cell membranes in confocal microscopy images. In: Burgos, N., Gooya, A., Svoboda, D. (eds.) SASHIMI 2019. LNCS, vol. 11827, pp. 81–89. Springer, Cham (2019). https://doi.org/10.1007/978-3-030-32778-1_9
4. Isola, P., Zhu, J.Y., Zhou, T., Efros, A.A.: Image-to-image translation with conditional adversarial networks. In: Proceedings of the IEEE Conference on Computer Vision and Pattern Recognition, pp. 1125–1134 (2017)
5. Karimi, D., Samei, G., Kesch, C., Nir, G., Salcudean, S.E.: Prostate segmentation in MRI using a convolutional neural network architecture and training strategy based on statistical shape models. Int. J. Comput. Assist. Radiol. Surg. **13**(8), 1211–1219 (2018)
6. Kingma, D.P., Welling, M.: Auto-encoding variational bayes. arXiv preprint arXiv:1312.6114 (2013)
7. Long, J., Shelhamer, E., Darrell, T.: Fully convolutional networks for semantic segmentation. In: Proceedings of the IEEE Conference on Computer Vision and Pattern Recognition, pp. 3431–3440 (2015)

8. Milletari, F., Rothberg, A., Jia, J., Sofka, M.: Integrating statistical prior knowledge into convolutional neural networks. In: Descoteaux, M., Maier-Hein, L., Franz, A., Jannin, P., Collins, D.L., Duchesne, S. (eds.) MICCAI 2017. LNCS, vol. 10433, pp. 161–168. Springer, Cham (2017). https://doi.org/10.1007/978-3-319-66182-7_19

9. Mitchell, S.C., Bosch, J.G., Lelieveldt, B.P., Van der Geest, R.J., Reiber, J.H., Sonka, M.: 3-D active appearance models: segmentation of cardiac MR and ultrasound images. IEEE Trans. Med. Imaging **21**(9), 1167–1178 (2002)

10. Ronneberger, O., Fischer, P., Brox, T.: U-Net: convolutional networks for biomedical image segmentation. In: Navab, N., Hornegger, J., Wells, W.M., Frangi, A.F. (eds.) MICCAI 2015. LNCS, vol. 9351, pp. 234–241. Springer, Cham (2015). https://doi.org/10.1007/978-3-319-24574-4_28

11. Tan, C., Yan, Z., Zhang, S., Li, K., Metaxas, D.N.: Collaborative multi-agent learning for MR knee articular cartilage segmentation. In: Shen, D., et al. (eds.) MICCAI 2019. LNCS, vol. 11765, pp. 282–290. Springer, Cham (2019). https://doi.org/10.1007/978-3-030-32245-8_32

12. Vercauteren, T., Pennec, X., Perchant, A., Ayache, N.: Diffeomorphic demons: efficient non-parametric image registration. NeuroImage **45**(1), S61–S72 (2009)

Interpretation of Brain Morphology in Association to Alzheimer's Disease Dementia Classification Using Graph Convolutional Networks on Triangulated Meshes

Emanuel A. Azcona[1,4]([✉])[iD], Pierre Besson[2,4][iD], Yunan Wu[1,4][iD], Arjun Punjabi[1,4][iD], Adam Martersteck[3,4][iD], Amil Dravid[1,4][iD], Todd B. Parrish[3,4][iD], S. Kathleen Bandt[2,4][iD], and Aggelos K. Katsaggelos[1,4][iD]

[1] Image and Video Processing Laboratory, Department of Electrical and Computer Engineering, Northwestern University, Evanston, IL, USA
emanuelazcona@u.northwestern.edu
[2] Advanced NeuroImaging and Surgical Epilepsy (ANISE) Lab, Northwestern Memorial Hospital, Chicago, IL, USA
[3] Neuroimaging Laboratory, Department of Radiology, Northwestern University, Evanston, IL, USA
[4] Augmented Intelligence in Medical Imaging, Northwestern University, Evanston, IL, USA
https://ivpl.northwestern.edu, https://anise-lab.com/
http://neuroimaging.northwestern.edu,

Abstract. We propose a mesh-based technique to aid in the classification of Alzheimer's disease dementia (ADD) using mesh representations of the cortex and subcortical structures. Deep learning methods for classification tasks that utilize structural neuroimaging often require extensive learning parameters to optimize. Frequently, these approaches for automated medical diagnosis also lack visual interpretability for areas in the brain involved in making a diagnosis. This work: (a) analyzes brain shape using surface information of the cortex and subcortical structures, (b) proposes a residual learning framework for state-of-the-art graph convolutional networks which offer a significant reduction in learnable parameters, and (c) offers visual interpretability of the network via class-specific gradient information that localizes important regions of interest in our inputs. With our proposed method leveraging the use of cortical and subcortical surface information, we outperform other machine learning methods with a 96.35% testing accuracy for the ADD vs. healthy control

Data used in preparation of this article were obtained from the Alzheimer's Disease Neuroimaging Initiative (ADNI) database (adni.loni.usc.edu). As such, the investigators within the ADNI contributed to the design and implementation of ADNI and/or provided data but did not participate in analysis or writing of this report. A complete listing of ADNI investigators can be found at: http://adni.loni.usc.edu/wp-content/uploads/how_to_apply/ADNI_Acknowledgement_List.pdf.

M. Reuter et al. (Eds.): ShapeMI 2020, LNCS 12474, pp. 95–107, 2020.
https://doi.org/10.1007/978-3-030-61056-2_8

problem. We confirm the validity of our model by observing its performance in a 25-trial Monte Carlo cross-validation. The generated visualization maps in our study show correspondences with current knowledge regarding the structural localization of pathological changes in the brain associated to dementia of the Alzheimer's type.

Keywords: Graph convolutional networks · Alzheimer's disease classification · Triangulated meshes · Neural network interpretability

1   Introduction

Alzheimer's disease dementia (ADD) is a clinical syndrome characterized by progressive amnestic multidomain cognitive impairment [27]. The causative underlying pathology is Alzheimer's disease (AD), defined as the co-occurrence of neurofibrillary tangles and amyloid-beta plaques. Globally, the number of individuals living with AD is expected to reach 1 out of 85 people by the year 2050 [4]. Automated methods for the computer-aided clinical diagnosis of ADD has been an area of interest in the medical imaging community for the development of assistive tools aiding in the visual inspection of structural information captured by magnetic resonance imaging (MRI).

Previous studies in the neuroanatomical pathologies of AD have demonstrated correlations in cortical folding pattern [5] and different neurodegenerative pathologies. Specific patterns of atrophy in the cortex and subcortical structures have been linked to AD [21,25]. For example, [5] discusses a potential to focus on high variability in association cortices like the intermediate sulcus of Jensen. As [28] also points out, widespread cortical thinning and a greater rate of atrophy is present in temporal lobe regions, primarily the left parahippocampal gyrus, for subjects with AD. Furthermore, Jong *et al.* [6] discuss irregularities like reduced putamen and thalamus volumes for subjects with AD. In studies such as ADNI, it is common to find bias towards more left-sided atrophy because of the verbal language tests given to assess memory function [8]. For example, if asymmetrical atrophy of the language network is more prominent, subjects may perform worse on verbal tests and be diagnosed with dementia earlier.

Machine learning (ML) methods have been a growing area of interest in the automated clinical diagnosis for ADD. [2,24,38] discuss the use of support vector machines (SVMs) in unimodal and multimodal imaging pipelines for the automated classification of ADD using MRI, PET, and cerebrospinal fluid (CSF). In [23,30], the use of MRI and PET imaging in multimodal convolutional neural networks (CNNs) for ADD diagnosis is discussed. SVM-based approaches, like those used in [2,24,38], have historically been hard to interpret, expensive to train, and often serve as the logical choice only when there is enough domain expertise to construct meaningful kernels. Multimodal volumetric CNNs like [30], often require a lot of memory and frequently are limited to smaller-batch operations or using lower resolution 3D volumes.

Motivated by 3D object detection via surfaces [26], cortical and subcortical irregularities correlated with ADD, our work uses mesh manifolds of the cortex and subcortical structures in the diagnosis of ADD. Our technique leverages a reduction in computational complexity offered by [7]. In [29], Parisot *et al.* leverage this work from [7] to make similar predictions for Alzheimer's disease and Autism using graph convolutional networks (GCNs) on ADNI/ABIDE subject population graphs. In [31], their convolutional mesh autoencoder (CoMA) framework uses the same GCN basis from [7] on human face surface meshes to generate new meshes from a learned distribution conditioned on facial expression labels. Their network is also able to reconstruct input meshes from compressed 8-dimensional representations with a 50% reduction in reconstruction error, while using 75% fewer parameters than volumetric models that operate on voxels.

The interpretability of results from ML models has remained an open issue in highlighting regions of interest (ROI) in relation to classification decisions. In this paper we demonstrate that it is possible to (1) extract meaningful surface meshes of the cortex and subcortical structures, (2) achieve accurate predictions for the clinical binary classification of ADD using meshes, (3) extract class-discriminative localization maps for interpretable ROI, and (4) reduce the number of learnable parameters.

2   Methods

Data used in the preparation of this article were obtained from the Alzheimer's Disease Neuroimaging Initiative (ADNI) database (https://adni.loni.usc.edu). The ADNI was launched in 2003 as a public-private partnership, led by Principal Investigator Michael W. Weiner, MD. The primary goal of ADNI has been to test whether serial magnetic resonance imaging (MRI), positron emission tomography (PET), other biological markers, and clinical and neuropsychological assessment can be combined to measure the progression of mild cognitive impairment (MCI) and early Alzheimer's disease (AD).

2.1   Localized Spectral Filtering on Graphs

Spectral-based graph convolution methods inherit ideas from a graph signal processing (GSP) perspective as described by [37]. Like [7], our work focuses on using undirected graphs defined by a finite set of vertices, $\mathcal{V}$, with $N = |\mathcal{V}|$ vertices, and a corresponding set of edges, $\mathcal{E}$, with scalar edge weights, $e_{ij} = e_{ji} \in \mathcal{E}$, which are stored in the ith rows and j^{th} columns of the adjacency matrix, $\mathbf{A} \in \mathbb{R}^{N \times N}$. A graph's node attributes are defined using the node feature matrix $\mathbf{X} \in \mathbb{R}^{N \times F}$ where each column, $\mathbf{x}_i \in \mathbb{R}^N$, represents the feature vector for a particular shared feature across each of the vertices, $v_i \in \mathcal{V}$.

A great emphasis in GSP is placed on the normalized graph Laplacian, $\mathbf{L} = \mathbf{I}_N - \mathbf{D}^{-1/2} \mathbf{A} \mathbf{D}^{-1/2}$, where $\mathbf{I}_N$ is the identity matrix and $\mathbf{D}_{ii} = \sum_j \mathbf{A}_{ij}$ is the diagonal matrix of node degrees. $\mathbf{L}$ can be factored via the eigendecomposition: $\mathbf{L} = \mathbf{U}\mathbf{\Lambda}\mathbf{U}^T$, where $\mathbf{U} \in \mathbb{R}^{N \times N}$ is the complete set of orthonormal eigenvectors for $\mathbf{L}$ and $\boldsymbol{\Lambda} = diag\left([\lambda_0, \ldots, \lambda_{N-1}]\right) \in \mathbb{R}^{N \times N}$ is the corresponding set of eigenvalues. Given a spectral filter, g_θ, defined in the graph's Fourier space [34] as a polynomial of the Laplacian, $\mathbf{L}$, and $\mathbf{U}$'s orthonormality, we can filter $\mathbf{x}$ via multiplication s.t.

$$g_\theta *_{\mathcal{G}} \mathbf{x} = g_\theta(\mathbf{L})\mathbf{x} = g_\theta\left(\mathbf{U}\boldsymbol{\Lambda}\mathbf{U}^T\right)\mathbf{x} = \mathbf{U}g_\theta(\boldsymbol{\Lambda})\mathbf{U}^T\mathbf{x}, \tag{1}$$

where $\theta \in \mathbb{R}^N$ are the parameters of the filter g_θ and $*_{\mathcal{G}}$ is the spectral convolution operator notation borrowed from [7]. Furthermore, $\mathbf{U}^T\mathbf{x}$ is the *graph Fourier transform* (GFT) of the graph signal $\mathbf{x}$, $g_\theta(\boldsymbol{\Lambda})$ is a filter defined using the spectrum (eigenvalues) of the normalized Laplacian, $\mathbf{L}$, and the left-sided multiplication with $\mathbf{U}$ is the *inverse*-GFT (IGFT). In this context, convolution is implicitly performed by using the duality property of the Fourier transform s.t. a spectral filter is first multiplied with the GFT of a signal, and then the IGFT of their product is determined.

Our approach uses Chebyshev polynomials of the first kind [1,7] to approximate g_θ using the graph's spectrum s.t.

$$g_\theta(\tilde{\mathbf{L}}) = \sum_{k=0}^{K-1} \theta_k T_k(\tilde{\mathbf{L}}), \tag{2}$$

for the scaled Laplacian $\tilde{\mathbf{L}} = \frac{2\mathbf{L}}{\lambda_{max}} - \mathbf{I}_N$, where λ_{max} is the largest eigenvalue in $\boldsymbol{\Lambda}$, and K can be interpreted as the kernel size. Chebyshev polynomials of the first kind are defined by the recurrence relation, $T_k(\tilde{\mathbf{L}}) = 2\tilde{\mathbf{L}}T_{k-1}(\tilde{\mathbf{L}}) - T_{k-2}(\tilde{\mathbf{L}})$ where $T_0(\tilde{\mathbf{L}}) = \mathbf{I}$ and $T_1(\tilde{\mathbf{L}}) = \tilde{\mathbf{L}}$ as shown in [7].

2.2   Mesh Extractions of Cortical and Subcortical Structures

Using FreeSurfer v6.0 [10], all MRIs were denoised followed by field inhomogeneity correction, and intensity and spatial normalization. Inner cortical surfaces (interface between gray and white matter) and outer cortical surfaces (CSF/gray matter interface) were extracted and automatically corrected for topological defects. Additionally, seven subcortical structures per hemisphere were segmented (amygdala, nucleus accumbens, caudate, hippocampus, pallidum, putamen, thalamus) and modeled into surface meshes using SPHARM-PDM (https://www.nitrc.org/projects/spharm-pdm).

Surfaces were inflated, parameterized to a sphere, and registered to a corresponding spherical surface template using a rigid-body registration to preserve the cortical [10] and subcortical [3] anatomy. Surface templates were converted to meshes using their triangulation schemes. A scalar edge weight, e_{ij}, was assigned to connect vertices v_i and v_j using their geodesic distance, ψ_{ij}, s.t.

$$e_{ij} = e_{ji} = \frac{1}{\sigma\sqrt{2\pi}} e^{-\frac{1}{2}\left(\frac{\psi_{ij}}{\sigma}\right)^2}. \tag{3}$$

Surface templates were parcellated using a hierarchical bipartite partitioning of their corresponding mesh. Starting with their initial mesh representation of densely triangulated surfaces, spectral clustering was used to define two partitions. These two groups were then each separated yielding four child partitions, and this process was repeated until the average distance across neighbor partitions was below 2.5 mm. For each partition, the central node was defined as the node whose centrality was highest and the distance across two partitions was defined as the geodesic distance (in mm) across the central vertices. Two partitions were neighbors if at least one node in each partition were connected. Finally, partitions were numbered so that partitions $2i$ and $2i+1$ at level L had the same parent partition i at level $L-1$. Therefore, for each level a graph was obtained s.t. the vertices of the graph were the central vertices of the partitions and the edges across neighboring vertices were weighted as in Eq. 3. This serves as an improvement upon [7] to ensure that no singleton is ever produced by pooling operations for the cortex and subcortical structures. At the finest level, meshes had a total of 47, 616 vertices: 32, 768 vertices for the cortex and 14, 848 vertices to represent the subcortical structures.

Vertex features were defined as the Cartesian coordinates of the surface vertices in the subjects' native space registered to the surface templates. This can create issues if the original scans are not registered to the same template, as was also done by Ranjan et al. in [31]. Similar studies, like that of Gutiérrez-Becker and Wachinger [15], implement "rotation network" modules as the first few layers of their neural network (NN) architecture to aid in correcting and aligning their samples to a common template. Performing our template registration as an additional preprocessing step reduces the complexity of our NN architecture and eliminates the need of incorporating an "alignment" term to our cost function to optimize later, as was needed in [15].

Cortical vertices were assigned 6 features: the x, y, and z coordinates of both the white matter (WM) and gray matter (GM) vertices in the native space. This was decided because vertices on these surfaces use the same edge weights and therefore the same "faces" with different coordinates for the vertices of the respective triangles. Similar to the cortex, subcortical vertices had 3 features: their corresponding x, y, and z coordinates in the native space as well. To maintain the same number of features for all vertices per scan, the corresponding cortical and subcortical feature matrices were block-diagonalized into a single node feature matrix per scan s.t. $\mathbf{X} \in \mathbb{R}^{47,616 \times 9}$. Sample meshes extracted from a randomly selected HC and one with ADD are demonstrated in Fig. 1.

2.3 Residual Network Architecture

Inspired by the work of He et al. in [16], we propose an improvement upon Cheb-Net [7] using residual connections within GCNs, which have been shown in prior work to address the common "vanishing gradient" problem and improve the performance of deep NNs. Typically, these types of residual networks (ResNets) are implemented by using batch normalization (BN) [19] before a ReLU activation function, and followed by convolution as seen in Fig. 2. Using ResBlocks (Fig. 2), max-pooling operations as described by Defferrard et al. [7], and a standard fully connected (FC) layer [32], the total architecture used in our study is defined in

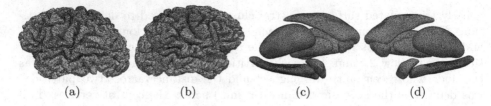

| (a) | (b) | (c) | (d) |

Fig. 1. Cortical meshes from a randomly selected HC subject (blue) and meshes of the subcortical structures from a randomly selected ADD subject (yellow). Presented are lateral views (a-b) of the HC's left hemisphere (LH) and right hemisphere (RH) cortical meshes respectively. Medial views of the ADD subject's LH and RH subcortical structure meshes are also presented (c-d). (Colour figure online)

Fig. 3. An additional ResBlock, which we refer to as a "post-ResBlock," was introduced prior to the FC layer as a linear mapping tool to match the number of FC units.

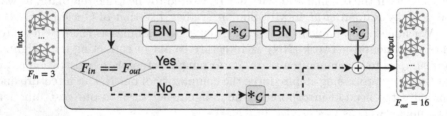

Fig. 2. Single ResBlock in the GCN architecture used in this study. Linear mapping of F_{in} to F_{out} channels is implemented using a convolutional layer, $*_{\mathcal{G}}$. This is done to match the number of input features to the number of desired feature maps.

2.4 Grad-CAM Mesh Adaptation

Interpretability of CNNs was addressed by [33] via their gradient-weighted class activation map (Grad-CAM) approach. In their work, images are fed to CNNs and gradients for each class score (logits prior to softmax) are extracted at the last convolutional layer. Using these gradients, they perform a global average pooling (GAP) operation for each feature map per class to extract "neuron importance weights," $\alpha_c^{(k)} \in \mathbb{R}^{c \times k}$, whose formulation we readapt for meshes s.t.

$$\alpha_c^{(k)} = \frac{1}{N} \sum_n \frac{\partial y_c}{\partial A_n^{(k)}}, \tag{4}$$

where y_c corresponds to the class score of class c, and $A_n^{(k)}$ refers to the value at vertex n for the k-th feature map $A^{(k)} \in \mathbb{R}^N$. A set of neuron importance

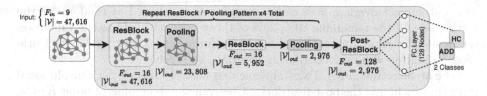

Fig. 3. Residual GCN used for the binary classification of ADD. In this study, max-pooling operations are used to downsample the vertex dimension by a factor of 2.

weights, $\alpha_c^{(k)}$, is extracted for each k-th feature map, $A^{(k)}$, and projected onto them to get the class activation maps (CAMs) s.t.

$$M_c = \mathrm{ReLU}\left(\sum_k \alpha_c^{(k)} A^{(k)}\right) \in \mathbb{R}^N. \tag{5}$$

As a consequence of pooling, CAMs are upsampled to the same number of nodes as the input mesh for a direct "overlay" using a trivial interpolation by going backward along the hierarchical tree used by the pooling operations.

3 Experimental Design

3.1 Dataset and Preprocessing

T1-weighted MRIs from ADNI [20] were selected with ADD/HC diagnosis labels given up to 2 months after the corresponding scan. This was taken as a precaution to ensure that each diagnosis had clinical justification. The dataset in our study consisted of 1,191 different scans for 435 unique subjects. Section 3.2 outlines our stratified data splitting strategy to ensure no data leakage occurs at the subject level across the training, validation, and testing sets [12].

Meshes for each MRI were extracted following the process described in Sect. 2.2. The spatial standard deviation from Eq. 3, σ, was set to 2 ad-hoc. The visual quality for each mesh was assessed manually via a direct overlay over slices of the corresponding MRI. Laplacians for the cortex and each subcortical structure were block-diagonalized to create one overall $\mathbf{L}$ representing a single mesh with multiple connected components. Extracted feature matrices for each sample were min-max normalized per feature to the interval $[-1, 1]$ prior to feeding batches of data into the networks. The added zeros during block-diagonalization (as discussed in Sect. 2.2) were ignored during each normalization step.

3.2 Network Architecture and Training

Extra care was taken in the shuffling of samples to avoid bias from subject overlap in our cross-validation [12]. A custom dataset splitting function was implemented s.t. the distribution of labels was preserved amongst each set while also ensuring to avoid subject overlap. 20% of the samples were selected at random for the testing set. Of the remaining 80%, 20% of those were withheld as the validation set, while the remaining belonged to the training set. A 25-trial Monte Carlo cross-validation was performed using this data split scheme.

The architecture in Fig. 3 was implemented using 16 kernels per convolutional layer (not including the post-ResBlock), Chebyshev polynomials of order $K = 3$, and pooling windows of size $p = 2$. Four alternating ResBlock and pooling layers were cascaded as shown in Fig. 3 prior to the post-ResBlock. The number of units at the post-ResBlock and FC layer was 128. Our GCN was optimized by minimizing a standard binary cross-entropy loss function

$$\mathcal{L} = -\frac{1}{N} \sum_{n=1}^{N} y_n \log(\hat{y_n}) + (1 - y_n) \log\left(1 - \hat{y_n}\right), \tag{6}$$

where $\hat{y_n}$ is the predicted class for the nth sample of N total samples and y_n is the ground truth label for the same sample index, n.

Networks were trained using batches of 32 samples per step for 100 epochs in each Monte Carlo trial. The Adam [22] optimizer was used with a learning rate of 5×10^{-4} and a learning rate decay of 0.999. Experiments were implemented in Python 3.6 using Tensorflow 1.13.4 using an NVIDIA GeForce GTX TITAN Z GPU in a Dell Precision Tower 7910 with Linux Mint 19.2.

4   Results and Discussion

4.1   ADD vs. HC Classification

Our cross-validation includes the same multilayer perceptron (MLP) classifier architecture, ridge classifier, and a 100-estimator random forest classifier set up by Parisot *et al.* in [29], where a similar graph approach is also used on the classification of ADD based on population graphs. The MLP designed was synonymous to the design in [29] s.t. the number of hidden layers and parameters was the same as our GCNs. Demonstrated in Fig. 4, our GCN outperformed other standard classifiers not limited to graph methods on our dataset split.

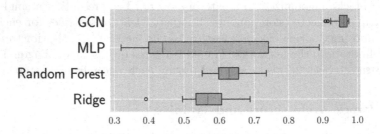

Fig. 4. Monte Carlo cross-validation accuracy results for GCN and baseline model architectures from [29] used on brain meshes.

Table 1. Model comparison to classifiers in studies not limited to surface methods.

Study	Data	ADD/HC	Acc. (%)	Sens. (%)	Spec. (%)	AUC (%)
[30]	MRI	$-/-$ (723)	73.76	–	–	–
[30]	MRI + PET$_{amyloid}$	$-/-$ (723)	92.34	–	–	–
[24]	MRI + PET$_{FDG}$	51/52	94.37	94.71	94.04	**97.24**
[36]	MRI + CSF	96/111	91.80	88.50	94.60	95.80
[18]	MRI	228/188	84.13	82.45	85.63	90.00
[2]	MRI	92/94	93.01	89.13	**96.80**	93.51
[17]	MRI	70/70	**97.60**	–	–	–
[14]	MRI	200/232	94.74	**95.24**	94.26	–
Ours	MRI	167/265	96.35	92.37	96.74	96.84

The results in Table 1 highlight comparable metrics of our model versus other studies that operate on voxels from full 3D MRI volumes, including [30]. In their work, Punjabi *et al.* train a multi-modal CNN using both volumetric MRI and FDG-PET imaging for the same task, which we outperform while training and evaluating on a smaller subset of their subject population. Furthermore, volumetric models like those in [30] rely on 3D CNNs with far more learned parameters, e.g.. [30]'s 200,194,502 weights ($\times$ 2 for fusion model), in comparison to our GCN's 497,522 learned parameters needed for comparable results. Like [31], we also achieve comparable results with far less learning parameters by working on meshes and focusing on brain *shape* instead of working on raw voxels obtained from MRIs and using voxel-based approaches.

4.2 Class Activation Map Visualization

By employing Grad-CAM on our best GCN, an average CAM was generated for true positive (TP) predictions (Fig. 5). We project our CAM onto the cortical template [11] provided by FreeSurfer [10] and the homemade subcortical structure templates detailed in [3]. The color scale highlights areas from least-to-most influential in TP predictions. The patterns in the CAM match previously described distributions of cortical and subcortical atrophy [9,21]. One reason we may observe a mismatch between the CAM and expected atrophy in the inferior parietal lobule could be the degree of variability in highly folded association cortex, e.g.., the intermediate sulcus of Jensen is found only in some individuals [5,35]. The slightly more left lateralized pattern in the CAM aligns with previous reports that propose greater pathologic burden and neurodegeneration of the language network which leads to worsening on verbal-based neuropsychological measures of memory resulting in a diagnosis for ADD [8].

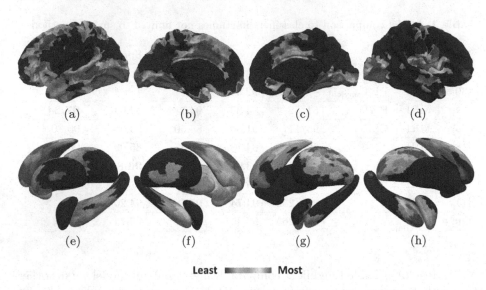

Fig. 5. Average TP CAMs on the cortical template from [10,11] (top) and subcortical structures from [3] (bottom) including: (a-b, e-f) lateral-medial views of the LH respectively, (c-d, g-h) medial-lateral views of the RH respectively.

5   Conclusion and Future Work

In this work, we demonstrated the effectiveness of using cortical and subcortical surface meshes in the context of binary ADD clinical diagnosis and ROI visualization in TP predictions. Furthermore, we compared the cross-validation results of our model for the same ADD vs. HC problem using other ML models on our data. Additionally, our final results were comparable to the results of other studies that use traditional neuroimaging modalities as inputs. When compared to the performance of the multimodal approach used in [30], our model outperforms their approach, thus potentially indicating the reliability of leveraging shape information represented as meshes to perform the same binary classification task.

Natural extensions of this work could be to (1) expand our classification problem to include a third class from ADNI, mild cognitive impairment (MCI), (2) increase the population in our study to include those in ADNI3 [20], (3) work on longitudinal predictions, and (4) compare our model's performance in using only the cortex, subcortical structures, or both. Additionally, having a 3D-volume-to-mesh dataset offers the potential for developing generative networks, as in [13], for performing the graph extraction preprocessing step described in Sect. 2.2. This will provide more autonomy and limit the need for the manual quality assessment (QA) of meshes as a part of our pipeline.

Acknowledgements. This work was funded in part by the Biomedical Data Driven Discovery Training Grant from the National Library of Medicine (5T32LM012203) through Northwestern University, and the National Institute on Aging. The authors would also like to thank the QUEST High Performance Computing Cluster at Northwestern University for computational resources.

Data collection and sharing for this project was funded by the Alzheimer's Disease Neuroimaging Initiative (ADNI). Data collection and sharing for this project was funded by the Alzheimer's Disease Neuroimaging Initiative (ADNI) (National Institutes of Health Grant U01 AG024904) and DOD ADNI (Department of Defense award number W81XWH-12-2-0012). ADNI is funded by the National Institute on Aging, the National Institute of Biomedical Imaging and Bioengineering, and through generous contributions from the following: AbbVie, Alzheimer's Association; Alzheimer's Drug Discovery Foundation; Araclon Biotech; BioClinica, Inc.; Biogen; Bristol-Myers Squibb Company; CereSpir, Inc.; Cogstate; Eisai Inc.; Elan Pharmaceuticals, Inc.; Eli Lilly and Company; EuroImmun; F. Hoffmann-La Roche Ltd. and its affiliated company Genentech, Inc.; Fujirebio; GE Healthcare; IXICO Ltd.; Janssen Alzheimer Immunotherapy Research & Development, LLC.; Johnson & Johnson Pharmaceutical Research & Development LLC.; Lumosity; Lundbeck; Merck & Co., Inc.; Meso Scale Diagnostics, LLC.; NeuroRx Research; Neurotrack Technologies; Novartis Pharmaceuticals Corporation; Pfizer Inc.; Piramal Imaging; Servier; Takeda Pharmaceutical Company; and Transition Therapeutics. The Canadian Institutes of Health Research is providing funds to support ADNI clinical sites in Canada. Private sector contributions are facilitated by the Foundation for the National Institutes of Health (https://www.fnih.org). The grantee organization is the Northern California Institute for Research and Education, and the study is coordinated by the Alzheimer's Therapeutic Research Institute at the University of Southern California. ADNI data are disseminated by the Laboratory for Neuro Imaging at the University of Southern California.

References

1. Arfken, G.B., Weber, H.J., Harris, F.E.: Mathematical Methods for Physicists. 3 edn. Academic Press, New York (2013). https://doi.org/10.1016/C2013-0-10310-8
2. Beheshti, I., et al.: Classification of Alzheimer's disease and prediction of mild cognitive impairment-to-Alzheimer's conversion from structural magnetic resource imaging using feature ranking and a genetic algorithm. Comput. Biol. Med. **83**, 109–119 (2017). https://doi.org/10.1016/j.compbiomed.2017.02.011
3. Besson, P., et al.: Intra-subject reliability of the high-resolution whole-brain structural connectome. NeuroImage **102**, 283–293 (2014). https://doi.org/10.1016/j.neuroimage.2014.07.064
4. Brookmeyer, R., et al.: Forecasting the global burden of Alzheimer's disease. Alzheimer Dement. **3**, 186–191 (2007). https://doi.org/10.1016/j.jalz.2007.04.381
5. Bruyn, G.: Atlas of the Cerebral Sulci, vol. 93. G. Thieme Verlag, New York (1991)
6. De Jong, L.W., et al.: Strongly reduced volumes of putamen and thalamus in Alzheimer's disease: an MRI study. Brain **131**(12), 3277–3285 (2008). https://doi.org/10.1093/brain/awn278
7. Defferrard, M., Bresson, X., Vandergheynst, P.: Convolutional neural networks on graphs with fast localized spectral filtering. In: Advances in Neural Information Processing Systems, pp. 3844–3852 (2016). http://papers.nips.cc/paper/6081-convolutional-neural-networks-on-graphs-with-fast-localized-spectral-filtering.pdf

8. Derflinger, S., et al.: Grey-matter atrophy in Alzheimer's disease is asymmetric but not lateralized. J. Alzheimer Dis. **25**(2), 347–357 (2011). https://doi.org/10.3233/JAD-2011-110041
9. Dickerson, B.C., et al.: The cortical signature of Alzheimer's disease: regionally specific cortical thinning relates to symptom severity in very mild to mild AD dementia and is detectable in asymptomatic amyloid-positive individuals. Cereb. Cortex **19**(3), 497–510 (2009). https://doi.org/10.1093/cercor/bhn113
10. Fischl, B.: FreeSurfer. NeuroImage **62**(2), 774–781 (2012). https://doi.org/10.1016/j.neuroimage.2012.01.021
11. Fischl, B., et al.: High-resolution intersubject averaging and a coordinate system for the cortical surface. Hum. Brain Mapp. **8**(4), 272–284 (1999). https://doi.org/10.1002/(SICI)1097-0193(1999)8:4⟨272::AID-HBM10⟩3.0.CO;2-4
12. Fung, Y.R., et al.: Alzheimer's disease brain MRI classification: challenges and insights. arXiv preprint arXiv:1906.04231 (2019). https://arxiv.org/abs/1906.04231
13. Goodfellow, I.J., et al.: Generative adversarial nets. In: Advances in Neural Information Processing Systems, vol. 3, pp. 2672–2680 (2014). http://papers.nips.cc/paper/5423-generative-adversarial-nets.pdf
14. Gupta, A., Ayhan, M., Maida, A.: Natural image bases to represent neuroimaging data. In: International Conference on Machine Learning, pp. 987–994 (2013). http://proceedings.mlr.press/v28/gupta13b.pdf
15. Gutiérrez-Becker, B., Wachinger, C.: Learning a conditional generative model for anatomical shape analysis. In: Chung, A.C.S., Gee, J.C., Yushkevich, P.A., Bao, S. (eds.) IPMI 2019. LNCS, vol. 11492, pp. 505–516. Springer, Cham (2019). https://doi.org/10.1007/978-3-030-20351-1_39
16. He, K., et al.: Deep residual learning for image recognition. In: Proceedings of the IEEE Computer Society Conference on Computer Vision and Pattern Recognition, 2016 December, pp. 770–778 (2016). https://doi.org/10.1109/CVPR.2016.90
17. Hosseini-Asl, E., Keynton, R., El-Baz, A.: Alzheimer's disease diagnostics by adaptation of 3d convolutional network. In: 2016 IEEE International Conference on Image Processing (ICIP), pp. 126–130. IEEE (2016). https://doi.org/10.1109/ICIP.2016.7532332
18. Hu, K., et al.: Multi-scale features extraction from baseline structure MRI for MCI patient classification and ad early diagnosis. Neurocomputing **175**, 132–145 (2016). https://doi.org/10.1016/j.neucom.2015.10.043
19. Ioffe, S., Szegedy, C.: Batch normalization: accelerating deep network training by reducing internal covariate shift. In: 32nd International Conference on Machine Learning, ICML 2015. Proceedings of Machine Learning Research, vol. 1, pp. 448–456 (2015). http://proceedings.mlr.press/v37/ioffe15.pdf
20. Jack, C.R., et al.: The Alzheimer's disease neuroimaging initiative (ADNI): MRI methods (2008). https://doi.org/10.1002/jmri.21049
21. Kälin, A.M., et al.: Subcortical shape changes, hippocampal atrophy and cortical thinning in future Alzheimer's disease patients. Front. Aging Neurosci. **9**, 38 (2017). https://doi.org/10.3389/fnagi.2017.00038
22. Kingma, D.P., Ba, J.L.: Adam: a method for stochastic optimization. In: 3rd International Conference on Learning Representations (2015). https://arxiv.org/pdf/1412.6980.pdf
23. Li, R., et al.: Deep learning based imaging data completion for improved brain disease diagnosis. In: Golland, P., Hata, N., Barillot, C., Hornegger, J., Howe, R. (eds.) MICCAI 2014. LNCS, vol. 8675, pp. 305–312. Springer, Cham (2014). https://doi.org/10.1007/978-3-319-10443-0_39

24. Liu, F., Wee, C.Y., Chen, H., Shen, D.: Inter-modality relationship constrained multi-modality multi-task feature selection for Alzheimer's disease and mild cognitive impairment identification. NeuroImage **84**, 466–475 (2014). https://doi.org/10.1016/j.neuroimage.2013.09.015

25. Liu, T., et al.: Cortical gyrification and sulcal spans in early stage Alzheimer's disease. PLoS ONE (2012). https://doi.org/10.1371/journal.pone.0031083

26. Masci, J., Boscaini, D., Bronstein, M.M., Vandergheynst, P.: Geodesic convolutional neural networks on riemannian manifolds. In: Proceedings of the IEEE International Conference on Computer Vision, 2015 February, pp. 832–840 (2015). https://doi.org/10.1109/ICCVW.2015.112

27. McKhann, G.M., et al.: The diagnosis of dementia due to Alzheimer's disease: Recommendations from the National Institute on Aging-Alzheimer's Association workgroups on diagnostic guidelines for Alzheimer's disease. Alzheimer Dement. **7**(3), 263–269 (2011). https://doi.org/10.1016/j.jalz.2011.03.005

28. Pacheco, J., et al.: Greater cortical thinning in normal older adults predicts later cognitive impairment. Neurobiol. Aging **36**(2), 903–908 (2015). https://doi.org/10.1016/j.neurobiolaging.2014.08.031

29. Parisot, S., et al.: Disease prediction using graph convolutional networks: application to autism spectrum disorder and Alzheimer's disease. Med. Image Anal. **48**, 117–130 (2018). https://doi.org/10.1016/j.media.2018.06.001

30. Punjabi, A., et al.: Neuroimaging modality fusion in Alzheimer's classification using convolutional neural networks. PLoS ONE **14**(12), 1–14 (2019). https://doi.org/10.1371/journal.pone.0225759

31. Ranjan, A., Bolkart, T., Sanyal, S., Black, M.J.: Generating 3D faces using convolutional mesh autoencoders. In: Ferrari, V., Hebert, M., Sminchisescu, C., Weiss, Y. (eds.) ECCV 2018. LNCS, vol. 11207, pp. 725–741. Springer, Cham (2018). https://doi.org/10.1007/978-3-030-01219-9_43

32. Rumelhart, D.E., Hinton, G.E., Williams, R.J.: Learning internal representations by error propagation. In: Readings in Cognitive Science: A Perspective from Psychology and Artificial Intelligence, pp. 399–421. MIT Press, Cambridge, MA (2013)

33. Selvaraju, R.R., et al.: Grad-CAM: visual explanations from deep networks via gradient-based localization. In: International Journal of Computer Vision (2019). https://doi.org/10.1109/ICCV.2017.74

34. Shuman, D.I., et al.: The emerging field of signal processing on graphs: extending high-dimensional data analysis to networks and other irregular domains. IEEE Signal Process. Mag. **30**(3), 83–98 (2013). https://doi.org/10.1109/MSP.2012.2235192

35. Thompson, P.M., et al.: Cortical variability and asymmetry in normal aging and Alzheimer's disease. Cereb. Cortex **8**(6), 492–509 (1998). https://doi.org/10.1093/cercor/8.6.492

36. Westman, E., Muehlboeck, J.S., Simmons, A.: Combining MRI and CSF measures for classification of Alzheimer's disease and prediction of mild cognitive impairment conversion. NeuroImage **62**(1), 229–238 (2012). https://doi.org/10.1016/j.neuroimage.2012.04.056

37. Wu, Z., et al.: A comprehensive survey on graph neural networks. IEEE Trans. Neural Netw. Learn. Syst. 1–21 (2020). https://doi.org/10.1109/TNNLS.2020.2978386

38. Zhang, D., Shen, D.: Multi-modal multi-task learning for joint prediction of multiple regression and classification variables in Alzheimer's disease. NeuroImage **59**(2), 895–907 (2012). https://doi.org/10.1016/j.neuroimage.2011.09.069

Applications

Combined Estimation of Shape and Pose for Statistical Analysis of Articulating Joints

Praful Agrawal[✉], Joseph D. Mozingo, Shireen Y. Elhabian,
Andrew E. Anderson, and Ross T. Whitaker

Scientific Computing and Imaging Institute, University of Utah, Salt Lake City, USA
prafulag@cs.utah.edu

Abstract. Quantifying shape variations in articulated joints is of utmost interest to understand the underlying joint biomechanics and associated clinical symptoms. For joint comparisons and analysis, the relative positions of the bones can confound subsequent analysis. Clinicians design specific image acquisition protocols to neutralize the individual pose variations. However, recent studies have shown that even specific acquisition protocols fail to achieve consistent pose. The individual pose variations are largely attributed to the day-to-day functioning of the patient, such as gait during walk, as well as interactions between specific morphologies and joint alignment. This paper presents a novel two-step method to neutralize such patient-specific variations while simultaneously preserving the inherent relationship of the articulated joint. The resulting shape models are then used to discover clinically relevant shape variations in a population of hip joints.

Keywords: Statistical shape modeling · Articulated joints · Pose alignment

1 Introduction

Statistical shape modeling of musculoskeletal structures enables a wide range of biomedical and clinical applications (e.g., [4, 8, 9, 16, 20, 22, 25, 26]). However, most studies are limited to isolated anatomies, and are rarely used to model articulated joints (e.g., hip and shoulder joints). Understanding the shape variations in articulated joints provides crucial insights into the associated biomechanics [23]. Further, these shape variations are often used to explain the clinically observed symptoms. For instance, the articulated relationship of femur and pelvis in the hip joint is particularly relevant in pathology that may lead to hip arthritis [3].

Statistical shape models aim to capture the subtle local variations (modulo of global geometric transformation) in a given anatomical population. For models that rely on explicit shape representation (i.e., correspondence models

© Springer Nature Switzerland AG 2020
M. Reuter et al. (Eds.): ShapeMI 2020, LNCS 12474, pp. 111–121, 2020.
https://doi.org/10.1007/978-3-030-61056-2_9

[10]), generalized Procrustes analysis (GPA) [19] is widely used to remove the similarity or rigid transformations before statistical analysis. However, the GPA method is not equipped to address variations in relative pose of distinct shapes, such as flexion and abduction, commonly observed in articulated joints. Clinical protocols impose specific image acquisition guidelines in an attempt to mitigate such variations [24]. Results from [24] show that even a consistent acquisition protocol may not completely remove the influence of pose. Therefore, a pose neutralization step is necessary before quantitative analysis of articulated joints. For this purpose, Bossa and Olmos [6] perform GPA on independent components of the multi-object shape models. This approach is further adopted in [7], [14], and [15] for functional modeling of articulated joints. The independent alignment of components helps address the undesired subject-specific pose variations; although at the expense of preserving information about relative position, some of which is robust to patient/subject pose. This trade-off undermines the validity of biomechanics-based functional modeling, where distances or geometric relationships among shapes is important. In another attempt, Gorczowski *et al.* [18] use a two-step approach, where the first step involves a rigid alignment of the joint shape models followed by the final step of component-wise GPA. Even though the initial rigid alignment preserves the relative position of the joint segments, it fails to neutralize the subject-specific scaling. Further, the component-wise GPA nullifies the important aspect of first step.

Bindernagel *et al.* [5] and Kainmueller *et al.* [21] devise alignment schemes for specific human joints. In both cases, a two-step alignment is performed to neutralize the individual pose. First the joint shape models are rigidly aligned, then a rigid alignment of one of the components is performed. This method relies on the clinical assumptions, i.e. consistent articulation, that may not be generalizable in all cases. Further, it is unclear how the model will evolve for more than two components. Audenaert *et al.* [2] use partial least squares (PLS) regression to estimate the shape model of articulated joint from independent shape models of the involved components. The output joint shape model is assumed free from the subject-specific pose. However, Audenaert *et al.* did not assume any model for the pose related *error* or a validation mechanism to support the claim.

Most existing methods rely on alignment of independent components of the joint, and forgo shape relationships. For the purpose of statistical shape analysis, it is important to factor out the undesired similarity transformation among samples in the population. However, the relative pose may be disease-specific and therefore, needs to be preserved for a clinically relevant modeling. With this goal, we propose a two-step alignment scheme, which removes the similarity transformation and individual pose variations from a population of articulated joints. The proposed method does not rely on selected template shape from the population. Further, the method is not limited to a particular human joint and is generalizable to more than two components. A population of dysplastic hip joints, involving pelvis and femur bones, is used for the experiments. Next, we discuss the details of proposed method, followed by a brief description of dataset and comparative approaches, and conclude with a discussion of experimental results.

2   Method

Consider a shape population of articulated joints $S = \{S_1, S_2, \cdots, S_N\}$, with K components i.e., $S_i = \{S_i^1, S_i^2, \cdots, S_i^K\}$. The goal is to quantify statistical shape variations among the sample joints in S. Statistical shape analysis relies on a correspondence model to consistently represent the sample shapes. Point-based models are a common choice for the purpose. In this section, we briefly discuss the method used to generate the point-based shape correspondence model, and describe the proposed two-step alignment scheme for a population of articulated joints.

2.1   Point-Based Shape Correspondence Model

Most existing methods optimize coordinate transformation between samples and a chosen template shape to generate the point-based correspondence models [2,14]. Therefore, the resulting analysis often varies based on the choice of template. A recent study [17] shows that the nonparametric, entropy-based method [12] outperforms the pairwise and template-based approaches. Geometric features can be incorporated within the entropy framework to address the complex anatomies, such as the scapula and pelvis [1,13]. Further, the correspondence model for multi-object complexes is produced by optimizing the statistics of combined shape [11]. For placing points on populations of shapes, we use the open-source implementation of this method, ShapeWorks [10].

2.2   Proposed Two-Step Alignment

It is imperative for statistical analysis to remove the global transformations (e.g., similarity or rigid) among samples in the shape population. Thus, GPA is used for analysis for the individual components across the population. In articulated joints, the statistical analysis aims for similar local anatomic variations in a simultaneous fashion across the components. However, it is important to neutralize the individual pose variations caused by flexion/extension, abduction/adduction, and internal/external rotation. In this regard, we propose the following two-step alignment:

1. *Global Alignment*: Consider each articulated joint S_i as a concatenated vector of all components $\{S_i^j\}_{j=1}^K$. GPA is performed on these joint shape vectors S_i, $i = 1, 2, \cdots, N$ to estimate the similarity transformations T_i^g, $i = 1, 2, \cdots, N$ resulting in aligned and size neutralized articulated joints $\widehat{S}_i = T_i^g(S_i)$. The GPA is an iterative method to estimate the transformation T_i^g between the sample shape S_i and a template shape S_μ such that

$$T_i^g = \underset{T}{\mathrm{argmin}} \; D^2(T(S_i), S_\mu), \qquad (1)$$

$$T^g(\mathbf{x}) = s\mathbf{R}\mathbf{x} + \mathbf{d}, \qquad (2)$$

where D^2 is the squared distance between the template and Procrustes aligned shape vectors. The average shape after every iteration is used as the template shape for the next. Using mean shape as the template eliminates bias toward a particular sample. The iterative alignment is performed until convergence. The average of these aligned joints now can be used as a template for individual alignments.

2. *Local Alignment*: The globally aligned joints $(\widehat{S}_i)$ still contain individual pose from the acquisition time. To achieve a consistent pose across samples while preserving the relative position of involved components, we use the average shape $\widehat{S_\mu} = \frac{1}{N}\sum_{i=1}^{N}\widehat{S}_i$ as the fixed template and perform rigid alignment of independent components. The component-wise transformations are estimated as

$$T_{ij}^l = \underset{T}{\mathrm{argmin}}\ D^2(T(\widehat{S_i^j}), \widehat{S_\mu^j}), \tag{3}$$

$$T^l(\mathbf{x}) = \mathbf{R}\mathbf{x} + \mathbf{d}. \tag{4}$$

The relative scale between components of individual joint is crucial for modeling the biomechanics and combined shape variations. Thus, we preserve the size of independent components after the global alignment step.

It is important to note that, if image acquisition protocols are enough to ensure a consistent relative position among the joints, the second step of Local Alignment should yield identity transformations. Similarly for the isolated anatomies, this method will be effectively reduced to only one step of global alignment. After the proposed alignment, each component in the joint shape vector $\widetilde{S}_i$ is computed as

$$\widetilde{S_i^j} = T_{ij}^l \circ T_i^g(S_i^j).$$

3   Experimental Setup

We used a dataset of 84 presurgery femur-pelvis pairs from 48 female patients with dysplasia[1]. The presence of dysplasia was confirmed based on the measurement of the lateral center edge angle (LCEA) acquired from a single antero-posterior radiograph. The age of the subjects ranged from 16 to 58 years, with a mean and standard deviation of 37.74 and 10.23 years, respectively. The CT acquisition protocol for patients with hip dysplasia specifies that they be in a supine position, legs straight with knee and toes pointing upward. The CT image resolution ranged from $(0.76 - 0.84) \times (0.76 - 0.84) \times (1 - 1.25)\,\mathrm{mm}^3$. The bones were segmented from CT images using Amira software (Visage Imaging, San Diego, CA, USA). The binary segmentations were then resampled to a consistent resolution of $0.5 \times 0.5 \times 0.5\,\mathrm{mm}^3$. The shape models with 2048 and 4096 correspondences on pelvis and femur, respectively were optimized using geometric features in ShapeWorks.

---

[1] Thanks to Dr. Tokunaga and Keisuke Uemura for data collection and providing this dataset.

Since the data was acquired using a consistent clinical protocol, we consider the shape model resulting from the global alignment step as a candidate model for comparison, this model was referred to as Global. As done in most existing methods, we generated comparative models using component-wise Procrustes alignment with the mean shape as the template. Two models (LocalSim and LocalRigid) were generated, where LocalSim considers similarity transformation (described in Eq. (2)) and LocalRigid uses rigid transformation (described in Eq. (4)). We used the mean shape and eigen analysis of joint shape vectors (S_i) to investigate the resulting statistical models.

4   Results and Discussion

Existing studies are often influenced by the choice of template for alignment. We chose to use mean shape as the template to eliminate sample bias. Therefore, as shown in Fig. 1, the mean shapes from four resulting models look nearly identical. A template free correspondence model generation also contributed to the consistent mean shapes. Further analysis of eigen spectrum and statistical modes of variation highlights the drawbacks in the local alignment and single step global alignment schemes. Since the articulated joint is disintegrated in both local alignment methods, we expect to see unnatural collision between the two components when we visualize the statistical modes of variation. As mentioned earlier, the global alignment is expected to fail to remove the rotation based pose variations. Next, we analyze the resulting eigen spectrum from the four methods.

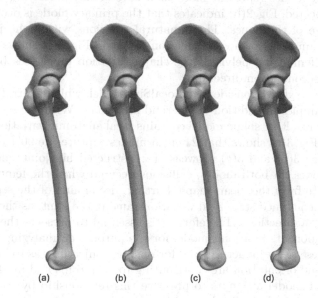

(a)   (b)   (c)   (d)

Fig. 1. Comparing mean shapes from the four models: (a) LocalRigid, (b) LocalSim, (c) Global, and (d) Proposed.

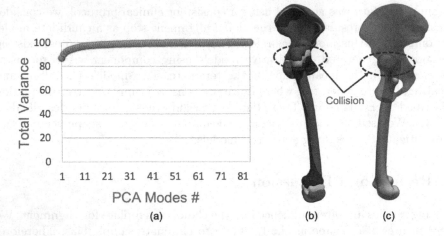

Fig. 2. LocalRigid: (a) Variance plot, (b) first eigen mode - yellow represents mean shape, blue and red are reconstructions for mean plus and minus two standard deviations, respectively, in the eigen direction, and (c) reconstructed shape representing the mean plus two standard deviations in second eigen direction.(Color figure online)

LocalRigid: Often in musculoskeletal shape analysis, scale is considered as a desired shape variation. Therefore, only rigid transformations are removed for the purpose of statistical modeling, and as is done when modeling joints [5,21]. Figure 2(a) shows the variance plot for the resulting statistical model. The top two dominant modes represent more than 90% of the underlying variation. As would be expected, Fig. 2(b) indicates that the primary mode is occupied by the change in size of the bones. The unnatural collisions within the joint renders this method unusable for pose estimation. Further, Fig. 2(c) shows the collision between the femur and pelvis as the shape variation progresses beyond mean towards the second eigen direction.

LocalSim: Next, we consider the LocalSim model, which neutralizes the scaling among samples in addition to the rigid alignment. With the scale variations removed, the resulting shape model contains local anatomic variations. The variance plot in Fig. 3(a) shows that 24 eigen modes capture the 90% of underlying variation. Figs. 3(b) and 3(c) showcase reconstructed hip joints along the first two eigen modes. In both modes, collisions occured when the femur and pelvis shapes deviate from the mean shape. Further, decoupling of the segments compromises the combined statistical variations among the joint, as the shapes vary along the eigen directions. Therefore, it is essential to preserve the relationship between components in a joint model for the purpose of analyzing the morphometrics of these tissue structures and for making conjectures as to how abnormal joint shape and orientation are contributing to symptoms and mechanical damage. The next model attempts to preserve this relationship by considering the entire joint as a single entity during alignment.

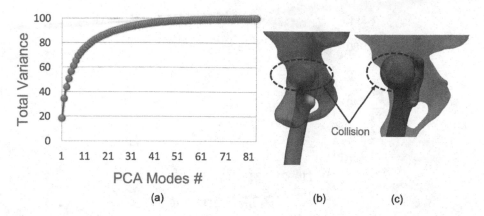

Fig. 3. LocalSim: (a) Variance plot, (b) reconstructed shape representing the mean minus two standard deviations in first eigen direction, and (c) reconstructed shape for the mean minus two standard deviations in second eigen direction.

Global: Here, we analyze the shape model after the first step of global alignment in the proposed two-step method. Figure 4(a) shows a much more compact variance plot compared to the LocalSim (Fig. 3(a)), with first six modes responsible for more than 90% of the population-wide statistics. This more compact variance plot indicates larger transformations compared to the targeted subtle anatomic variations. Figures 4(b)–4(d) validate this hypothesis, as we observe the articulated rotations clinically described as flexion/extension, abduction/adduction, and medial rotation. It is important to note that the order of these rotations in the eigen spectrum closely matches with the possible range of movement in hip joint. This crucial finding further highlights the need of pose neutralization at the level of shape modeling, regardless of a consistent image acquisition protocol.

Proposed: Using the average joint shape from global alignment as the template helps to preserve the relative position within the joint. Thus, we do not observe any unnatural shape variation as the femur and pelvis jointly vary along the eigen modes. A total of 18 modes represent the 90% of shape variations; Fig. 5(a) shows the cumulative variance plot. The first eigen mode in Fig. 5(b) shows the importance of relative size of components in the joint. The shape variation jointly captures the twist along femoral shaft in conjunction with the simultaneous change in the size of the pelvis and femoral head shapes. This combined behaviour results in variable femoral head coverage, a clinically relevant feature to diagnose dysplasia [24]. Second mode in Fig. 5(c) captures the synchronized twisting of the femur and pelvis, with femur twist along the shaft and acetabulum being the center of twist in pelvis. Clinicians expect such variations and often attribute the variation in coverage to this combined shape variation. Figure 5(d) shows some more subtle twist as the two shape vary along the third eigen mode.

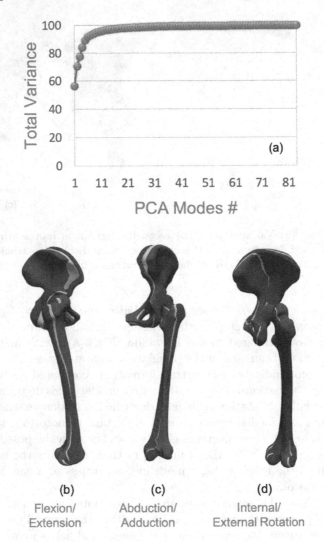

(a)

(b)
Flexion/
Extension

(c)
Abduction/
Adduction

(d)
Internal/
External Rotation

Fig. 4. Global: (a) Variance plot, (b) first eigen mode representing the flexion and extension in hip joint, (c) second eigen mode representing the abduction and adduction, and (d) fourth eigen mode representing the internal and external rotation. (b–d) Yellow represents mean shape, blue and red are reconstructions for mean plus and minus two standard deviations in the respective eigen directions.(Color figure online)

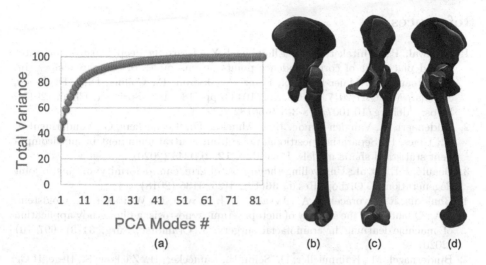

Fig. 5. Proposed method: (a) Variance plot, (b) first eigen mode, (c) second eigen mode, and (d) third eigen mode. (b–d) Yellow represents mean shape, blue and red are reconstructions for mean plus and minus two standard deviations in the respective eigen directions.(Color figure online)

5   Conclusion

This paper addresses a crucial problem of pose neutralization in the analysis of articulated joints. Using a point-based statistical shape model, and the proposed two-step alignment we are able to discover the clinically relevant shape variations in a population of patients with hip dysplasia. The described approach has the potential to influence clinical diagnostics and clinical decision making by quantifying variation in both the shape of a joint and its orientation in space. Comparisons between the various models described herein demonstrates the limitations of prior methods to address pose variations. Further, the statistical modes of shape variation show that clinical practices of a consistent acquisition fail to remove individual rotations. The proposed method is not limited to the hip anatomy and is applicable to other joints that may have two or more segments.

Funding Acknowledgments. This work was supported by the National Institutes of Health under grant numbers NIBIB-U24EB029011, NIAMS-R01AR076120, NHLBI-R01HL135568, NIBIB-R01EB016701, and NIGMS-P41GM103545. The content is solely the responsibility of the authors and does not necessarily represent the official views of the National Institutes of Health.

References

1. Agrawal, P., Whitaker, R.T., Elhabian, S.Y.: Learning deep features for automated placement of correspondence points on ensembles of complex shapes. In: Descoteaux, M., Maier-Hein, L., Franz, A., Jannin, P., Collins, D.L., Duchesne, S. (eds.) MICCAI 2017. LNCS, vol. 10433, pp. 185–193. Springer, Cham (2017). https://doi.org/10.1007/978-3-319-66182-7_22
2. Audenaert, E., Van den Eynde, J., de Almeida, D., Steenackers, G., Vandermeulen, D., Claes, P.: Separating positional noise from neutral alignment in multicomponent statistical shape models. Bone Rep. **12**, 100243 (2020)
3. Beaulé, P.E., et al.: Unravelling the hip pistol grip/cam deformity: origins to joint degeneration. J. Orthop. Res.® **36**(12), 3125–3135 (2018)
4. Bhalodia, R., Dvoracek, L.A., Ayyash, A.M., Kavan, L., Whitaker, R., Goldstein, J.A.: Quantifying the severity of metopic craniosynostosis: a pilot study application of machine learning in craniofacial surgery. J. Craniofac. Surg. **31**(3), 697–701 (2020)
5. Bindernagel, M., Kainmueller, D., Seim, H., Lamecker, H., Zachow, S., Hege, H.C.: An articulated statistical shape model of the human knee. In: Bildverarbeitung für die Medizin 2011, pp. 59–63. Springer, Heidelberg (2011). https://doi.org/10.1007/978-3-642-19335-4_14
6. Bossa, M.N., Olmos, S.: Statistical model of similarity transformations: building a multi-object pose. In: CVPRW 2006, p. 59. IEEE (2006)
7. Bossa, M.N., Olmos, S.: Multi-object statistical pose+shape models. In: ISBI, pp. 1204–1207 (2007)
8. Bryan, R., Nair, P.B., Taylor, M.: Use of a statistical model of the whole femur in a large scale, multi-model study of femoral neck fracture risk. J. Biomech. **42**(13), 2171–2176 (2009)
9. Cates, J., et al.: Computational shape models characterize shape change of the left atrium in atrial fibrillation. Clin. Med. Insights Cardiol. **8**, 99–109 (2014). CMC-S15710
10. Cates, J., Elhabian, S., Whitaker, R.: Shapeworks: particle-based shape correspondence and visualization software. In: Statistical Shape and Deformation Analysis, pp. 257–298. Elsevier (2017)
11. Cates, J., Fletcher, P.T., Styner, M., Hazlett, H.C., Whitaker, R.: Particle-based shape analysis of multi-object complexes. In: Metaxas, D., Axel, L., Fichtinger, G., Székely, G. (eds.) MICCAI 2008. LNCS, vol. 5241, pp. 477–485. Springer, Heidelberg (2008). https://doi.org/10.1007/978-3-540-85988-8_57
12. Cates, J., Fletcher, P.T., Styner, M., Shenton, M., Whitaker, R.: Shape modeling and analysis with entropy-based particle systems. In: Karssemeijer, N., Lelieveldt, B. (eds.) IPMI 2007. LNCS, vol. 4584, pp. 333–345. Springer, Heidelberg (2007). https://doi.org/10.1007/978-3-540-73273-0_28
13. Datar, M., Lyu, I., Kim, S.H., Cates, J., Styner, M.A., Whitaker, R.: Geodesic distances to landmarks for dense correspondence on ensembles of complex shapes. In: Mori, K., Sakuma, I., Sato, Y., Barillot, C., Navab, N. (eds.) MICCAI 2013. LNCS, vol. 8150, pp. 19–26. Springer, Heidelberg (2013). https://doi.org/10.1007/978-3-642-40763-5_3
14. Fouefack, J.R., Alemneh, T., Borotikar, B., Burdin, V., Douglas, T.S., Mutsvangwa, T.: Statistical shape-kinematics models of the skeletal joints: application to the shoulder complex. In: EMBC, pp. 4815–4818 (2019)

15. Fouefack, J.R., Borotikar, B., Douglas, T.S., Burdin, V., Mutsvangwa, T.E.M.: Dynamic multi-object Gaussian process models: a framework for data-driven functional modelling of human joints (2020)

16. Galloway, F., et al.: A large scale finite element study of a cementless osseointegrated tibial tray. J. Biomech. **46**(11), 1900–1906 (2013)

17. Goparaju, A., et al.: On the evaluation and validation of off-the-shelf statistical shape modeling tools: a clinical application. In: Reuter, M., et al. (eds.) ShapeMI 2018. LNCS, vol. 11167, pp. 14–27. Springer, Cham (2018). https://doi.org/10.1007/978-3-030-04747-4_2

18. Gorczowski, K., et al.: Statistical shape analysis of multi-object complexes. In: 2007 IEEE Conference on Computer Vision and Pattern Recognition, pp. 1–8. IEEE (2007)

19. Gower, J.C.: Generalized procrustes analysis. Psychometrika **40**(1), 33–51 (1975). https://doi.org/10.1007/BF02291478

20. Harris, M.D., Datar, M., Whitaker, R.T., Jurrus, E.R., Peters, C.L., Anderson, A.E.: Statistical shape modeling of cam femoroacetabular impingement. J. Orthop. Res. **31**(10), 1620–1626 (2013)

21. Kainmueller, D., Lamecker, H., Zachow, S., Hege, H.C.: An articulated statistical shape model for accurate hip joint segmentation. In: 2009 Annual International Conference of the IEEE Engineering in Medicine and Biology Society, pp. 6345–6351. IEEE (2009)

22. Kozic, N., et al.: Optimisation of orthopaedic implant design using statistical shape space analysis based on level sets. Med. Image Anal. **14**(3), 265–275 (2010)

23. Smoger, L.M.: Statistical modeling to characterize relationships between knee anatomy and kinematics. J. Orthop. Res.® **33**(11), 1620–1630 (2015)

24. Uemura, K., Atkins, P.R., Maas, S.A., Peters, C.L., Anderson, A.E.: Three-dimensional femoral head coverage in the standing position represents that measured in vivo during gait. Clin. Anat. **31**(8), 1177–1183 (2018)

25. Wang, D., Shi, L., Griffith, J.F., Qin, L., Yew, D.T., Riggs, C.M.: Comprehensive surface-based morphometry reveals the association of fracture risk and bone geometry. J. Orthop. Res. **30**(8), 1277–1284 (2012)

26. Zhao, Z., Taylor, W.D., Styner, M., Steffens, D.C., Krishnan, K.R.R., MacFall, J.R.: Hippocampus shape analysis and late-life depression. PLoS ONE **3**(3), e1837 (2008)

Learning a Statistical Full Spine Model from Partial Observations

Di Meng[1]([✉]), Marilyn Keller[2], Edmond Boyer[1], Michael Black[2], and Sergi Pujades[1]

[1] Inria, Univ. Grenoble Alpes, CNRS, Grenoble INP, LJK, Grenoble, France
di.meng@inria.fr
[2] Max Planck Institute for Intelligent Systems, Tübingen, Germany

Abstract. The study of the morphology of the human spine has attracted research attention for its many potential applications, such as image segmentation, bio-mechanics or pathology detection. However, as of today there is no publicly available statistical model of the 3D surface of the full spine. This is mainly due to the lack of openly available 3D data where the full spine is imaged and segmented. In this paper we propose to learn a statistical surface model of the full-spine (7 cervical, 12 thoracic and 5 lumbar vertebrae) from partial and incomplete views of the spine. In order to deal with the partial observations we use probabilistic principal component analysis (PPCA) to learn a surface shape model of the full spine. Quantitative evaluation demonstrates that the obtained model faithfully captures the shape of the population in a low dimensional space and generalizes to left out data. Furthermore, we show that the model faithfully captures the global correlations among the vertebrae shape. Given a partial observation of the spine, i.e. a few vertebrae, the model can predict the shape of unseen vertebrae with a mean error under 3 mm. The full-spine statistical model is trained on the VerSe 2019 public dataset and is publicly made available to the community for non-commercial purposes. (https://gitlab.inria.fr/spine/spine_model)

Keywords: Spine statistical model · Vertebrae surface · Incomplete data

1   Introduction

A reliable spine model with accurate vertebrae structures is essential in numerous medical application such as image segmentation, orthopedics, anesthesiology and pathology quantification. A full spine model can help segment 3D volumetric images such as computed tomography (CT) or magnetic resonance imaging (MRI), as well as 2D projections like X-ray images, for instance by injecting prior knowledge on the global shape of the spine as a regularizer. Statistical models have also shown their interest in the generation of synthetic data for different modalities. A full spine model can also be used as a reference structure in clinical

© Springer Nature Switzerland AG 2020
M. Reuter et al. (Eds.): ShapeMI 2020, LNCS 12474, pp. 122–133, 2020.
https://doi.org/10.1007/978-3-030-61056-2_10

practice supporting the localization of other organs, as well as be used in the diagnosis of spine scoliosis.

The statistical shape model is one of the most employed statistical atlas [10]. Creating statistical models of shape using surface measurements has attracted a lot of attention since the early work of Blanz and Vetter [2]. While early methods mostly rely on Principal Component Analysis, non-linear methods have been also proposed to overcome the fact that unseen shapes cannot be expressed by linear combinations of the training samples [24].

In order to create a statistical surface model of the full spine using the classical methodology, the full spine of several subjects needs to be observed. However, observations of the full spine (e.g. CT images) are rare. Practitioners usually narrow down the field of view of the spine to one specific level for a detailed diagnosis as well as to reduce the dose given to the patient. Thus most work has been done to model the shape of the vertebrae which are consistently observed in the scans. For instance, lumbar vertebrae have received a special attention for low back pain as this area supports the greatest load of the spine [5]. Hollenbeck et al. [11] generated a statistical shape model of the lumbar region from 52 subjects using PCA with a focus on the L4-L5 and L5-S1 functional spinal units (FSU). Campbell et al. [6] used automated methods for landmark identification to create a statistical shape model of the lumbar spine, applied in biomechanics and population-based finite element modeling. A nonlinear SSM based on kernel PCA was investigated for 3D active shape model segmentation [15]. They used 7 CT scans covering vertebrae level from T10 to L3.

Another difficulty in the learning of statistical models of the spine is that the relative position of vertebrae highly depends on the patient's posture during data acquisition. Overlapping and excessive separation of neighboring structures may appear in the vertebrae segmentation. In order to tackle this problem, a model of the interspace of the vertebrae was developed [8], by considering the variations of the space between two surfaces. Similarly, our model only focuses on the relative translation of the vertebrae and disregards their relative rotation. Multi-vertebrae statistical models [4, 19] capture the variations in shape and pose simultaneously, and reduce the number of registration parameters, allowing to help in the segmentation of a section of the vertebral column. These models were tested on 32 subjects and with a focus on the lumbar vertebrae section.

While most works have focused on the modeling of a specific part of the spine, a few works have modeled the full spine. For instance, Klinder et al. [13] designed an automatic framework segmenting vertebrae from arbitrary CT images with a full spine surface model. To create the model they first scanned a commercially available plastic phantom to create the template, and then they manually registered it to ten actual scans of the full spine. Mirzaalian et al. [17] learned a statistical shape model of the spine by independently learning three models, one for each level (cervical, thoracic and lumbar). Thus, their per-level models do not learn the shape correlations across the full spine.

Other probabilistic models, different from the shape surface models, such as probabilistic atlas [20], graph models [23], hidden markov models [9] and hierarchical models [7,22,26] have also been proposed. For example Ruiz et al. [20] proposed a probabilistic atlas of the spine. By co-registering 21 CT scans, a probability map is created which can be used to segment and detect the vertebrae with a special focus on ribs suppression. Schmidt et al. [23] proposed a probabilistic graphical model for the location and identification of the vertebrae in MR images. In both cases full spines were observed at train time and the proposed methods cannot be used to infer the shape of the full spine from a partial observation.

As full spines are not publicly available, we learn our novel statistical surface model from partial spine observations using Probabilistic Principal Component Analysis (PPCA). In the medical domain, PPCA has been used for craniofacial statistical modeling [16]. However, in their work full observations of the skull were used for training. In contrast, our proposed model never observed a complete spine. A robust shape model can capture the characteristics of individual vertebrae and the anatomy of the spine. It also provides the appearance information and relative positions of the observed objects. However, existing statistical models have focused on a specific region of the spine or were learned from full spine observations. Unfortunately, none of them are publicly available.

Our full spine statistical model is composed of 24 vertebrae including cervical, thoracic and lumbar. The model is learned from the openly available VerSe 2019 Dataset [21] containing partial observations and annotations of the spine. Our model captures the strong shape correlations across the individual vertebrae of the spine. Given an arbitrary field-of-view and incomplete scan, our model is able to predict the full spine adapting to the shape variations of the subject. The model is openly available to the community for non-commercial purposes.

2   Method

The inputs to our method are i) 80 manually segmented CT volumes $\mathbf{V}^i$, where $i \in [1, 80]$ denotes the volume index, together with the annotations of the individual vertebra from the VerSe Dataset [21] and ii) 24 artist created mesh templates $\mathbf{T}_k$ with $k \in [1, .., 24]$ being the vertebra index, one for each individual vertebra. Each template $\mathbf{T}_k$ has N_k 3D vertices and the full spine mesh including all 24 vertebrae has 127.294 vertices and 254.636 faces. All template meshes $\mathbf{T}_k$ are symmetric with respect to the medial plane. From the segmented volumes $\mathbf{V}^i$ we extract meshes using the Marching Cubes approach. We name them scans and note them as $\mathcal{S}_k^i$. Let us note that: i) in the VerSe Dataset none of the volumes contains the 24 vertebra; ii) some vertebra are imaged at the boundary of the CT volume and thus only provide an incomplete observation. These scans have holes and are not watertight; we note them $\bar{\mathcal{S}}_k^i$. An overview of the VerSe Dataset vertebrae is shown in Fig. 1.

The creation of the statistical spine shape model consists of several steps. First, in Sect. 2.1 we detail how we perform a non-rigid registration of each template $\mathbf{T}_k$ to all watertight scans $\{\mathcal{S}_k^i - \bar{\mathcal{S}}_k^i\}$. To bring all registrations to a coherent frame, we compute a rigid transformation [1] between the registration and the template. We name these coherent registrations $\mathcal{I}_k^i$. Then, for each vertebra k, we compute Principal Component Analysis (PCA) on all the available registrations. As we have few observations, we exploit the medial plane symmetry: for each registration we create the symmetric registration. With the registrations and their symmetric we obtain 24 (symmetric) mean shapes $\mathbf{T}_k^\mu$ and their 24 individual principal vectors $\mathbf{B}_k$. With the obtained PCA, we perform a second non-rigid registration of each template $\mathbf{T}_k$ as explained in Sect. 2.2. This time all scans $\mathcal{S}_k^i$, including the non-watertight, are registered. We initialize the template k vertices with the mean shape $\mathbf{T}_k^\mu$ and use the computed shape space $\mathbf{B}_k$ to constrain the registration process. We compute the rigid transformation to obtain the coherent registrations and keep the corresponding rotation R_k^i and translation t_k^i. We exploit again the medial plane symmetry and obtain a new set of registrations $\mathcal{A}_k^i$. From these registrations, we use Probabilistic Principal Component Analysis (PPCA) (Sect. 2.3) and we learn the full-spine shape model (Sect. 2.4) defined by $\mathbf{T}^\mu$ and $\mathbf{B}$.

2.1 Initial Registration

We first perform a non-rigid registration of each template $\mathbf{T}_k$ to all watertight scans $\{\mathcal{S}_k^i - \bar{\mathcal{S}}_k^i\}$. Because the scans are watertight, we want all template vertices to be close to the scan surface, and vice versa, all scan vertices to be close to the template surface. We effectively enforce this constraint by computing the vertex to surface distance from a point set to a surface and define the energy $E_{p2m}(\mathcal{S}, \mathbf{T})$, which accounts for the distance of the vertices of $\mathcal{S}$ to the mesh surface $\mathbf{T}$. The registration has three steps. Translation optimization, translation and rotation optimization and free form optimization. To simplify notation we drop the volume indices i and keep k to indicate the per-vertebrae process.

We start by computing a translation $\mathbf{t}$, so that $\mathbf{T}_k + \mathbf{t}$ is close to $\mathcal{S}$ by minimizing

$$E(\mathcal{S}, \mathbf{T}_k, \mathbf{t}) = E_{p2m}(\mathcal{S}, \mathbf{T}_k + \mathbf{t}) + \lambda_{m2s}E_{p2m}(\mathbf{T}_k + \mathbf{t}, \mathcal{S}) \tag{1}$$

w.r.t. to $\mathbf{t}$, where $\lambda_{m2s} = 1$. Next we optimize for a 3D rotation, parametrized by a 3D Rodriguez vector $\mathbf{r}$. Given a 3D vector $\mathbf{r}$ and a mesh $\mathbf{T}$, we use $R(\mathbf{T}, \mathbf{r})$ to describe the 3D rotated mesh. The CT scans from the VerSe Dataset (see Sect. 3) have a consistent patient orientation encoded in the dicom metadata. After reorienting all the CTs along the same orientation, we can thus initialize the rotation vector r ($r = [-2, 0, 2]$) with the same value for all $i \in [1, 80]$. This value corresponds to the rotation between the template frame and the acquisition frame defined by the radiologist. Then we optimize

$$E(\mathcal{S}, \mathbf{T}_k, \mathbf{t}, \mathbf{r}) = E_{p2m}(\mathcal{S}, R(\mathbf{T}_k, \mathbf{r}) + \mathbf{t}) + \lambda_{m2s}E_{p2m}(R(\mathbf{T}_k, \mathbf{r}) + \mathbf{t}, \mathcal{S}) \tag{2}$$

w.r.t. to $\mathbf{t}$ and $\mathbf{r}$, where $\mathbf{t}$ is initialized with the result of (1). Next we allow all vertices of $R(\mathbf{T}_k, \mathbf{r}) + \mathbf{t}$ to freely deform to best match $\mathcal{S}_k^i$. This free-form deformation is represented as an additive 3D vector added to each vertex that we note $\mathbf{f}$. To regularize the position of these displacements we use a coupling term on edges $E_{cpl}(\mathbf{T}, \mathbf{T} + \mathbf{f})$, enforcing the edges of the registration to be close to the edges of the initial template shape. To compute E_{cpl} we use the same energy term defined in Eq. 8 from [3] and optimize

$$
\begin{aligned}
E(\mathcal{S}, \mathbf{T}_k, \mathbf{t}, \mathbf{r}, \mathbf{f}) = {} & E_{p2m}(\mathcal{S}, (R(\mathbf{T}_k, \mathbf{r}) + \mathbf{t}) + \mathbf{f}) \\
& + \lambda_{m2s} E_{p2m}((R(\mathbf{T}_k, \mathbf{r}) + \mathbf{t}) + \mathbf{f}, \mathcal{S}) \\
& + \lambda_{cpl} E_{cpl}((R(\mathbf{T}_k, \mathbf{r}) + \mathbf{t}), (R(\mathbf{T}_k, \mathbf{r}) + \mathbf{t}) + \mathbf{f}) \quad (3)
\end{aligned}
$$

w.r.t. to $\mathbf{f}$ and by keeping $\mathbf{t}$ and $\mathbf{r}$ fixed. The weight λ_{cpl} was empirically chosen to leverage the contribution of each term into the final energy, namely $\lambda_{cpl} = 0.01$ for all experiments. Then we compute a rigid transformation [1] between the obtained registration and the template and obtain $\mathcal{I}_k^i$. From these registrations we compute a per vertebrae mean template $\mathbf{T}_k^\mu$ and perform PCA. We note by $\mathbf{B}_k$ the 10 first PCA principal components of vertebra k.

2.2 PCA-Guided Registration

In the PCA-guided registration we first compute $\mathbf{t}$ and $\mathbf{r}$ for non-watertight scans $\bar{\mathcal{S}}_k^i$, by optimizing (1) and (2). We set $\lambda_{m2s} = 0$ as not every vertex in the template must explain a scan point. Then, we use the individual shape space to parametrize a mesh as $\mathbf{T}_k^\mu + \beta \mathbf{B}_k$ where β is a 10-dimensional vector. We solve for $\mathbf{t}$, $\mathbf{r}$ and β to minimize

$$
\begin{aligned}
E(\mathcal{S}, \mathbf{T}_k^\mu, \mathbf{t}, \mathbf{r}, \beta) = {} & E_{p2m}(\mathcal{S}, R(\mathbf{T}_k^\mu, \mathbf{r}) + \mathbf{t} + \beta \mathbf{B}_k) \\
& + \lambda_{m2s} E_{p2m}(R(\mathbf{T}_k^\mu, \mathbf{r}) + \mathbf{t} + \beta \mathbf{B}_k, \mathcal{S}) \quad (4)
\end{aligned}
$$

with $\lambda_{m2s} = 1$ if the scan is watertight and $\lambda_{m2s} = 0$ otherwise. Then we perform a free vertices optimization following (3) by coupling the vertices to the solution obtained from (4). We compute the rigid transformation [1] between the obtained registration and the mean template $\mathbf{T}_k^\mu$ and obtain the final unposed registrations $\mathcal{A}_k^i$. These registrations will be used to learn the full spine model.

2.3 PPCA on the Registrations

For each volume index i, we do not have the full set of registration $\mathcal{A}_k^i$, as some k are not observed in the volume i (see Fig. 1). Thus we use Probabilistic Principal Component Analysis (PPCA) [25], a variant of PCA dealing with missing data. We construct a matrix with size $N \times S$ values, where $N = \sum_{k=1}^{24} 3N_k$, and $S = 160$ is the number of volumes used (80 volumes plus their symmetric version). Each column i is the concatenation of the 24 vectors, each given by the registration $\mathcal{A}_k^i$ vertices, if it is available, or an empty vector if the data is missing. The created matrix has 57% missing values. We compute PPCA using the publicly available implementation in [18] and obtain the mean spine $\mathbf{T}^\mu$ and the associated shape space $\mathbf{B}$.

2.4   The Spine Model

The architecture of the spine model is a graphical model mesh architecture parametrized by the shape parameters and the pose parameters, where each vertebrae has its own translation and rotation. The shape parameters β apply additive offsets to the initial vertebrae mesh $\mathbf{T}^\mu$. The offsets are the PCA principal components learned with the PPCA and the poses and translations are rigidly applied to each vertebra. To constrain the locations of the vertebrae, we learn a distribution over the relative positions of neighboring vertebrae. For each pair of neighboring vertebrae k and $k+1$, we use R_k^i, t_k^i and $t_{(k+1)}^i$ to compute the relative translation of vertebra $k+1$ w.r.t. vertebra k. Then for each 23 pairs $k, k+1$ we fit a Gaussian model to the observed relative translations. These relative translations distribution accounts for the thickness of the inter-vertebral disk. However, as the relative rotation between vertebrae highly depends on the patient posture, we do not learn a pose prior on the relative rotations.

3   VerSe Challenge Dataset

To train our model we use the publicly available MICCAI Challenge 2019 VerSe dataset [21]. It provides 80 CT scans with manually annotated voxel-wise labels, and 40 CT scans without labels. We use the 80 annotated subjects to create the spine model. Most volumes have a voxel size $1 \times 1 \times 1\,\mathrm{mm}^3$ and a few $1 \times 1 \times 2\,\mathrm{mm}^3$. We resample all volumes to have a fixed $1 \times 1 \times 1\,\mathrm{mm}^3$ sampling. An overview of the available observations from the full spine in the VerSe datset is shown in Fig. 1. Each row in the figure represents a vertebrae, C1 to T5 from top to bottom. Each column represents a CT volume. As we can see, each CT scan only covers a few vertebrae and no full spine is observed in any CT volume. In addition, the number of volumes including a given vertebra is not balanced across different vertebrae. The dataset is dominated by lumbar observations and few cervical vertebrae are observed. The section including the junction between cervical and thoracic vertebrae is critically sparse.

In our work we aim at learning the shape correlations across the full spine and to evaluate our model we use cross validation. In order to make sure that the junction between the cervical and thoracic vertebrae is always seen at train time, we label a volume *junction volume* if C5 to T2 are observed in the volume, and *not junction volume* otherwise. We use this junction label to create a balanced cross validation 8-fold setting for all our experiments.

As our method is trained on a few samples, we carefully inspected the input data. We noticed some misannotations in the ground truth labels. For example, some misannotations appear as floating points far away from the corresponding vertebra, as well as lumbar volumes having cervical labels. We cleaned the ground truth meshes by keeping only the biggest connected component in the mesh extracted from the CT volume. This procedure is standard as the evaluation criteria of the VerSe dataset only considers the biggest component in the volume [21]. We also observed a few vertebrae containing metal implants. As their shape is severely biased, we manually removed them from the dataset.

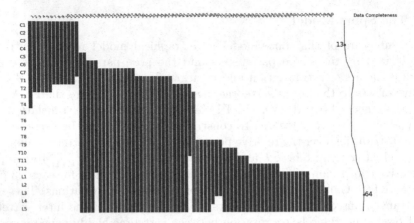

Fig. 1. Overview of the VerSe dataset. Rows represent one of the 24 individual vertebrae (C1 to L5). Column are the observed patients. No full spine is observed for any patient. The columns in red are the test set of one of the 8-fold used for cross validation.

4   Experiments

To demonstrate the accuracy and benefits from the spine model we perform two evaluations. In a first set of experiments we assess the metric accuracy of the registrations, the created model and the compacity of the learned shape space. In a second evaluation we assess the accuracy of our model to predict the shape of missing vertebrae.

4.1   Metric Accuracy

Before we learn the spine model, we first show that the registrations closely capture the input data shape with sub-voxel accuracy. Then we measure the compacity and the generalization power of the full spine model.

Individual Vertebrae Registrations Accuracy. To assess the accuracy of the individual registrations obtained in Sect. 2.2 we compute the point to mesh distance between the scans S_k^i and their registrations A_k^i. For each individual vertebra we aggregate the mean distance per scan by taking the mean over all available volumes. We report the mean error and standard deviation for each vertebra in Table 1. We aggregate the data by weighting each vertebrae errors with the number of samples and 0.24 mm errors in mean 0.20 mm std. Let us recall that the input voxel size is $1 \times 1 \times 1$ mm^3, thus the registrations faithfully capture the input data shape with a sub-voxel accuracy.

Full-Spine Model Accuracy. To assess the compacity of the learned spine shape space we show on the left of Fig. 2 the cumulative variance of the shape space. The fist 10 shape components capture 85% of the variance, and 90% variance

Table 1. Vertebrae registration error. me: mean error. std: standard deviation (mm).

	me	std		me	std		me	std		me	std		me	std		me	std
C1	0.20	0.16	**C2**	0.20	0.16	**C3**	0.19	0.15	**C4**	0.19	0.15	**C5**	0.19	0.15	**C6**	0.18	0.15
C7	0.21	0.16	**T1**	0.21	0.17	**T2**	0.21	0.17	**T3**	0.21	0.16	**T4**	0.21	0.16	**T5**	0.21	0.17
T6	0.21	0.17	**T7**	0.22	0.19	**T8**	0.23	0.19	**T9**	0.23	0.20	**T10**	0.23	0.20	**T11**	0.25	0.22
T12	0.25	0.21	**L1**	0.25	0.21	**L2**	0.26	0.22	**L3**	0.27	0.23	**L4**	0.28	0.25	**L5**	0.29	0.26

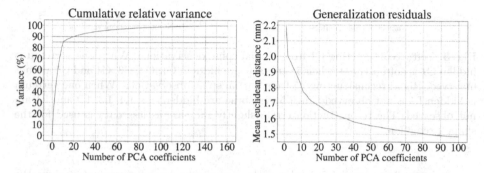

Fig. 2. Left: the cumulative variance of the full-spine model wrt the number of used shape components. Right: the generalization error of the model (cross validation) wrt to the number of used shape components.

is captured by the first 20. Let us recall that each registration has 127.294 vertices and 254.636 faces. To assess how well the model can capture the shape of unseen spines, we perform an 8-fold cross validation (see Sect. 3). We learn the shape space with a train set and evaluate it on the unseen test set. For an unseen, potentially partial, registration, we complete the missing vertebrae by using the learned PPCA, we then project it into the PCA space and reconstruct it with a selected number of components. We then measure the vertex to vertex error between the registration and the reconstruction. Only original data, i.e. not completed by the PPCA, is used in the metrics. On the right of Fig. 2 the generalization errors w.r.t. the number of used components are presented. Using 10 components we obtain a mean reconstruction error of 1.80 mm, and using 20 components we obtain a mean error of 1.68 mm.

4.2 Missing Vertebrae Shape Prediction Accuracy

Next, we aim at quantifying how well the model captures the global correlations across the spine. With the learned model, given a partial observation of the spine we can reconstruct the missing spine shape. We setup the following experiment to measure the accuracy of the shape of the predicted vertebrae.

For a cross validation fold, we learn the spine model on the train test. Then we take all registrations obtained in Sect. 2.2 in the test fold containing the cervical vertebrae C1 to C6. We first mask one vertebra starting from C1 and reconstruct it using the rest of the volume vertebrae. We then remove two vertebrae (C1 and C2) and reconstruct them using the rest. Progressively we remove

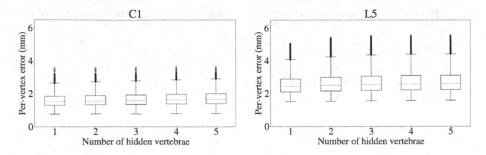

Fig. 3. Left: reconstructing C1 from partial spine observations. When one vertebra is hidden C1 is not observed, when 5 vertebrae are hidden, C1, C2, C3, C4 and C5 are not observed. Right: reconstructing L5 from partial spine observations. When one vertebra is hidden L5 is not observed, when 5 vertebrae are hidden, L5, L4, L3, L2 and L1 are not observed. The reconstruction errors slightly increase as less data is used for the shape prediction.

cervical vertebrae until C5, and use the remaining ones for reconstruction. We measure the vertex to vertex distance between the reconstructed vertebrae and the original registration. In an analogous way, we perform the same experiment for lumbar vertebrae, by progressively removing L5 to L1.

We use the 8-fold cross validation scheme (see Sect. 3) and evaluate on all volumes for which the data is available. In Fig. 3 we report the per vertex mean error over all volumes for C1 and L5. In the left plot, the reconstruction errors on C1, when masking 1, 2, 3, 4 and 5 vertebrae are reported. In the right plot, the results for the L5 reconstruction experiment are presented. In Fig. 4 we present the per vertex mean errors on the mean template.

We aggregate the per vertex errors for all C1 and obtain reconstruction errors 1.61 mm when one vertebra is masked, 1.65 mm when two vertebrae are masked, 1.68 mm when three vertebrae are masked, 1.72 mm when four vertebrae are masked and 1.75 mm when five vertebrae are masked. For the L5 vertebrae we obtain 2.57 mm when one vertebra is masked, 2.65 mm when two vertebrae are masked, 2.70 mm when three vertebrae are masked, 2.73 mm when four vertebrae are masked and 2.74 mm when five vertebrae are masked.

A trend in the increase of the error is observed as more vertebrae are masked, but the errors do not drastically increase. This trend can be seen in Fig. 3 as well as in the aggregated errors. These experiments show that the shape of all vertebrae is strongly correlated throughout the spine, and by observing a subset of the spine, the learned model can infer a plausible shape for the missing vertebrae.

1 left out   2 left out   3 left out   4 left out   5 left out

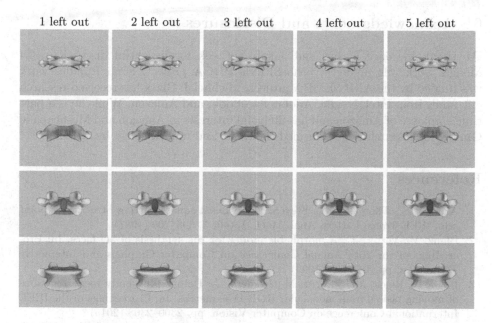

Fig. 4. Per vertex mean reconstruction error when N ∈ [1,2,3,4,5] vertebrae are left out, visualized on the mean shape. First two rows: C1 reconstructions front and rear views. Red is 4 mm. Blue is 0 mm. Third and fourth rows: L5 reconstructions front and rear views. Red is 6 mm. Blue is 0 mm. Errors slightly increase as less vertebrae are used for the reconstruction.(Color figure online)

5   Conclusion

In this paper we present a statistical surface model of the full-spine learned solely from partial and incomplete views of the spine. In order to deal with the partial observations we use probabilistic principal component analysis (PPCA) to learn a surface shape model of the full spine. The nature of the spine, specifically the correlations across the vertebrae shape, make the use of PPCA particularly well-suited. The quantitative evaluation demonstrates that the obtained model faithfully captures the shape of the population in a low dimensional space and generalizes to left out data. The model is openly available to the community for non-commercial purposes at https://gitlab.inria.fr/spine/spine_model.

While in this paper we presented an application of the model to predict the shape of missing vertebrae, future work will address the application of the model in other tasks, such as the generation of synthetic data, its registration to 2D X-ray images and its use in the segmentation and identification of vertebrae in volumetric images.

6   Acknowledgments and Disclosures

We thank Florence Forbes and Jakob Verbeek for insightful discussions. Di Meng's work was funded by the SPINE PDCA project. Sergi Pujades work was funded by the ANR SEMBA project. Michael J. Black has received research gift funds from Intel, Nvidia, Adobe, Facebook, and Amazon. While he is a part time employee of Amazon and has financial interests in Amazon and Meshcapade GmbH, his research was performed solely at, and funded solely by, MPI.

References

1. Arun, K.S., Thomas, S.H., Steven, D.B.: Least-squares fitting of two 3-D point sets. IEEE Trans. Pattern Anal. Mach. Intell. 5, 698–700 (1987)
2. Blanz, V., Vetter, T.: A morphable model for the synthesis of 3D faces. In: Proceedings of the 26th Annual Conference on Computer Graphics and Interactive Techniques, pp. 187–194 (1999)
3. Bogo, F., Black, M.J., Loper, M., Romero, J.: Detailed full-body reconstructions of moving people from monocular RGB-D sequences. In: Proceedings of the IEEE International Conference on Computer Vision, pp. 2300–2308 (2015)
4. Bossa, M.N., Olmos, S.: Multi-object statistical pose+ shape models. In: 2007 4th IEEE International Symposium on Biomedical Imaging: From Nano to Macro, pp. 1204–1207. IEEE (2007)
5. Bruners, P., et al.: Electromagnetic tracking for CT-guided spine interventions: phantom, ex-vivo and in-vivo results. Eur. Radiol. 19(4), 990–994 (2009). https:// doi.org/10.1007/s00330-008-1227-z
6. Campbell, J.Q., Petrella, A.J.: Automated finite element modeling of the lumbar spine: using a statistical shape model to generate a virtual population of models. J. Biomech. 49(13), 2593–2599 (2016)
7. Cai, Y., Osman, S., Sharma, M., Landis, M., Li, S.: Multi-modality vertebra recognition in arbitrary views using 3D deformable hierarchical model. IEEE Trans. Med. Imaging 34(8), 1676–1693 (2015)
8. Castro-Mateos, I., Pozo, J.M., Pereañez, M., Lekadir, K., Lazary, A., Frangi, A.F.: Statistical interspace models (SIMs): application to robust 3D spine segmentation. IEEE Trans. Med. Imaging 34(8), 1663–1675 (2015)
9. Glocker, B., Feulner, J., Criminisi, A., Haynor, D.R., Konukoglu, E.: Automatic localization and identification of vertebrae in arbitrary field-of-view CT scans. In: Ayache, N., Delingette, H., Golland, P., Mori, K. (eds.) MICCAI 2012. LNCS, vol. 7512, pp. 590–598. Springer, Heidelberg (2012). https://doi.org/10.1007/978-3-642-33454-2_73
10. Heimann, T., Meinzer, H.-P.: Statistical shape models for 3D medical image segmentation: a review. Med. Image Anal. 13(4), 543–563 (2009)
11. Hollenbeck, J.F.M., Cain, C.M., Fattor, J.A., Rullkoetter, P.J., Laz, P.J.: Statistical shape modeling characterizes three-dimensional shape and alignment variability in the lumbar spine. J. Biomech. 69, 146–155 (2018)
12. Kadoury, S., Labelle, H., Paragios, N.: Automatic inference of articulated spine models in CT images using high-order Markov random fields. Med. Image Anal. 15(4), 426–437 (2011)

13. Klinder, T., Ostermann, J., Ehm, M., Franz, A., Kneser, R., Lorenz, C.: Automated model-based vertebra detection, identification, and segmentation in CT images. Med. Image Anal. **13**(3), 471–482 (2009)
14. Korez, R., Ibragimov, B., Likar, B., Pernuš, F., Vrtovec, T.: A framework for automated spine and vertebrae interpolation-based detection and model-based segmentation. IEEE Trans. Med. Imaging **34**(8), 1649–1662 (2015)
15. Kirschner, M., Becker, M., Wesarg, S.: 3D active shape model segmentation with nonlinear shape priors. In: Fichtinger, G., Martel, A., Peters, T. (eds.) MICCAI 2011. LNCS, vol. 6892, pp. 492–499. Springer, Heidelberg (2011). https://doi.org/10.1007/978-3-642-23629-7_60
16. Lüthi, M., Albrecht, T., Vetter, T.: Building shape models from lousy data. In: Yang, G.-Z., Hawkes, D., Rueckert, D., Noble, A., Taylor, C. (eds.) MICCAI 2009. LNCS, vol. 5762, pp. 1–8. Springer, Heidelberg (2009). https://doi.org/10.1007/978-3-642-04271-3_1
17. Mirzaalian, H., Wels, M., Heimann, T., Kelm, B.M., Suehling. M.: Fast and robust 3D vertebra segmentation using statistical shape models. In: 2013 35th Annual International Conference of the IEEE Engineering in Medicine and Biology Society (EMBC), pp. 3379–3382. IEEE (2013)
18. Probabilistic PCA python implementation. https://github.com/shergreen/pyppca
19. Rasoulian, A., Rohling, R., Abolmaesumi, P.: Lumbar spine segmentation using a statistical multi-vertebrae anatomical shape+ pose model. IEEE Trans. Med. Imaging **32**(10), 1890–1900 (2013)
20. Ruiz-España, S., Domingo, J., Díaz-Parra, A., Dura, E., D'Ocón-Alcañiz, V., Arana, E., Moratal, D.: Automatic segmentation of the spine by means of a probabilistic atlas with a special focus on ribs suppression. Med. Phys. **44**(9), 4695–4707 (2017)
21. Sekuboyina, et al.: VerSe: A Vertebrae Labelling and Segmentation Benchmark. arXiv eprint: 2001.09193. arXiv:2001.09193 (2020)
22. Seifert, S.: Hierarchical parsing and semantic navigation of full body CT data. In: Medical Imaging 2009: Image Processing. International Society for Optics and Photonics (2009)
23. Schmidt, S., Kappes, J., Bergtholdt, M., Pekar, V., Dries, S., Bystrov, D., Schnörr, C.: Spine detection and labeling using a parts-based graphical model. In: Karssemeijer, N., Lelieveldt, B. (eds.) IPMI 2007. LNCS, vol. 4584, pp. 122–133. Springer, Heidelberg (2007). https://doi.org/10.1007/978-3-540-73273-0_11
24. Stacklies, W., Redestig, H., Scholz, M., Walther, D., Selbig, J.: pcaMethods–a bioconductor package providing PCA methods for incomplete data. Bioinformatics **23**(9), 1164–1167 (2007)
25. Tipping, M.E., Bishop, C.M.: Probabilistic principal component analysis. J. Roy. Stat. Soc.: Ser. B (Stat. Methodol.) **61**(3), 611–622 (1999)
26. Zhan, Y., Maneesh, D., Harder, M., Zhou, X.S.: Robust MR spine detection using hierarchical learning and local articulated model. In: Ayache, N., Delingette, H., Golland, P., Mori, K. (eds.) MICCAI 2012. LNCS, vol. 7510, pp. 141–148. Springer, Heidelberg (2012). https://doi.org/10.1007/978-3-642-33415-3_18

Morphology-Based Individual Vertebrae Classification

Eslam Mohammed, Di Meng[(✉)], and Sergi Pujades

Inria, Univ. Grenoble Alpes, CNRS, Grenoble INP, LJK, Grenoble, France
di.meng@inria.fr

Abstract. The human spine is composed, in non-pathological cases, of 24 vertebrae. Most vertebrae are morphologically distinct from the others, such as C1 (Atlas) or C2 (Axis), but some are morphologically closer, such as neighboring thoracic or lumbar vertebrae. In this work, we aim at quantifying to which extent the shape of a single vertebra is discriminating. We use a publicly available MICCAI VerSe 2019 Challenge dataset containing individually segmented vertebrae from CT images. We train several variants of a baseline 3D convolutional neural network (CNN) taking a binary volumetric representation of an isolated vertebra as input and regressing the vertebra class. We start by predicting the probability of the vertebrae to belong to each of the 24 classes. Then we study a second approach based on a two-stage aggregated classification which first identifies the anatomic group (cervical, thoracic or lumbar) then uses a group-specific network for the individual classification.

Our results show that: i) the shape of an individual vertebra can be used to faithfully identify its group (cervical, thoracic or lumbar), ii) the shape of the cervical and lumbar seems to have enough information for a reliable individual identification, and iii) the thoracic vertebrae seem to have the highest similarity and are the ones where the network is confused the most. Future work will study if other representations (such as meshes or pointclouds) obtain similar results, i.e. does the representation have an impact in the prediction accuracy?

Keywords: Vertebrae identification · Vertebrae group classification

1 Introduction

The human spine is usually composed of 24 vertebrae. They are structured in three anatomic groups: seven cervical (C1-C7), twelve thoracic (T1-T12) and five lumbar (L1-L5). Each group shares morphological and functional characteristics motivating their anatomic group classification. The three groups are illustrated in Fig. 1. Automatic identification of vertebrae in spinal imaging, such as Computed Tomography (CT) or Magnetic Resonance Imaging (MRI), is crucial in the context of clinical diagnosis and surgical planning. While some vertebrae, such as the first cervical (C1) have a very distinctive shape, other vertebrae, such as

© Springer Nature Switzerland AG 2020
M. Reuter et al. (Eds.): ShapeMI 2020, LNCS 12474, pp. 134–144, 2020.
https://doi.org/10.1007/978-3-030-61056-2_11

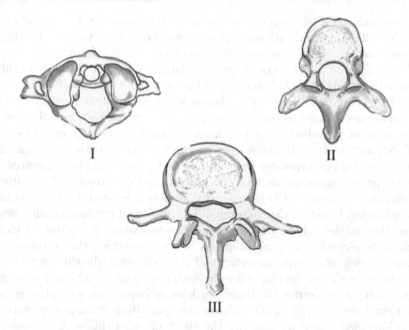

Fig. 1. Illustration of the shape of the vertebrae [10]. I - Representative shape of cervical C3-C7. II - Representative shape of the thoracic T1-T12. III - Representative shape of the lumbar L1-L5.

neighboring thoracic or lumbar vertebrae share a visually similar morphological appearance.

In this work, we aim at quantifying to which extent the shape of a single vertebra is discriminating. This knowledge aims at providing a source of relevant information to the global full spine identification methods. For example, distinctive vertebrae can be used as reliable anchor points in global methods, whereas the contribution of unreliable predictions can be diminished. As of today, the classification of all vertebrae in a CT volume still remains a major challenge for the community [14].

In this work we propose to train a 3D convolutional neural network (CNN) architecture that, given the morphology of an isolated vertebra as input, can first predict the probability of the vertebrae to belong to each of the 24 classes. Then, motivated by the fact that the anatomic groups have distinctive features, as illustrated in Fig. 1, we study a second approach by first identifying the anatomic group (cervical, thoracic or lumbar), and then, its individual identification with a per-group specialized network. To perform our study, we use the publicly available MICCAI VerSe 2019 Challenge dataset [14]. As the medical dataset contains a small number of annotated samples, we study the impact of several augmentation techniques (rotation, translation, scaling, noise addition) in the classification task. Preliminary results show that: i) the shape of an individual vertebra can be used to faithfully identify its group (cervical, thoracic or lumbar), ii) the shape of the cervical and lumbar seems to have enough information for a reliable individual identification, and iii) the thoracic vertebrae seem to have the highest similarity and are the ones where the network is confused the most.

Many computer-aided tasks in medical imaging were classically done via feature based methods [1,7]. AlexNet [4] was the first to obtain remarkable results in many of the visual challenges (e.g. classification, segmentation and detection). Henceforth, researchers in medical imaging have further studied the use and improvement of convolutional neural networks for a wide range of applications. The U-Net architecture [13] came up with an architecture for segmenting microscopic images that incorporates two main paths for capturing image context and other for precise localization. The 3D multi-task fully connected architecture [6] describes a complex yet comprehensive approach for vertebral segmentation and localization based on the contextual information of the surrounding organs. Indeed, this implies relying on an accurate methodology for vertebral identification. More recently, the work proposed by [5] segmenting the vertebrae in an iterative manner, states the importance of the individual identification of vertebrae. The network can decide, using the instance memory, whether to segment the next vertebra or retrieve it. In our work we re-implement a baseline convolutional neural network [15] in 3D, to learn and quantify how accurately vertebrae can be identified. Also, we quantify the effect of using different augmentation techniques in the prediction accuracy of the trained models.

2   Methodology

In our work, we start by extracting and pre-processing the vertebrae data, then we augment it with different strategies. With the processed data as input we train several networks to predict the individual vertebrae classification.

2.1   Data Extraction

The dataset used for our study is the VerSe dataset [14], which is a spine dataset of 80 CT scans with voxel-level vertebral annotations. The challenge of the dataset is to achieve segmentation, identification and localization of the vertebrae in CT volumes. In our work we use the ground truth annotation masks of the volumes as input data to train the classifiers. We extract every individual vertebra from the CT volumes and we obtain binary volumes (masks) for each vertebra.

Connected Components. The annotation masks of the VerSe dataset contain noise in the form of small isolated groups of voxels. After extracting the individual vertebrae, we use a 3D *connected component* algorithm to extract the biggest connected component(e.g. our desired vertebra). Then, we use the bounding box of the biggest connected component to obtain the clean volume.

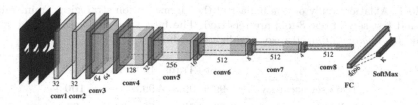

Fig. 2. The proposed 3D Convolutional Neural Network. Numbers under each block describe the size of the output kernels after each operation. Numbers on the z-axis of each block describe the size of the output cube in R^3.

Padding. Volumetric networks take as input a predefined volume structure having the same size. However, the obtained volumes from the previous step have different dimensions. To decide on the network input size we use the largest volume dimension in the dataset plus a margin of approx. 20%, namely a cube of size $128 \times 128 \times 128$ - where a power of two was preferred. All volumes are then centered and zero-padded to match this size.

2.2 Augmentation

A common useful technique to overcome the scarcity of data while training neural networks is *data augmentation*. Proposing variations of the same instance is essential to teach the network the desired invariance and robustness when there are few data samples for training. That is, adding more samples to leverage the training process by learning more complex features that consolidate the network's discriminative ability towards more robust features. We used four augmentation techniques: rotations, translations, scale and additive noise.

We considered the *rotations* around the 3 axes, hence introducing multiple orientations to the network. Because the acquired volumes have a coherent global orientation, we uniformly sampled angles θ in the range of $[-20, 20]^3$ degrees.

In order to teach the networks the *translation invariance* we feed as training samples translated inputs of the desired objects. So, instead of learning a centered object in the cube, samples are shifted by a $\delta \in R^3$ offset. The translation is uniformly sampled from the interval $[-20, 20]^3$ mm.

Another transformation that we consider is the *scaling* property of an object. We applied a uniform scaling factor $\gamma \in R$, uniformly sampling from the interval $[0.8, 1.2]$

We also added *additive noise* to the input data. We used *salt-and-pepper* noise by sampling from a *Poisson distribution* with parameter $\lambda = 0.05$.

Our mechanism to generate an augmented dataset involves applying 10 random combinations of the aforementioned transformations to each bone. Hence, we have 10 random versions of each bone as training samples. All these transformations were conducted while preserving the input size cube to $128 \times 128 \times 128$.

Table 1. Ablation study on the impact of the augmentation strategies. Train, validation and test sets of one 8-fold are reported. The best value is consistently obtained for the *full augmentation* strategy, highlighted in bold.

	Rot	Trans	Rot+Trans	Full
Train set accuracy	99.5	99.8	99.8	**100**
Validation set accuracy	73	70	75	**81**
Test set accuracy	74	67	70	**80**

2.3   Network Architecture

Our CNN architecture described in Fig. 2 is composed of 8 convolutional blocks followed by a *fully connected* layer 4096. Then a *SoftMax* layer is applied to get the probabilities of each class of the input volumes. Each convolutional block (except for the early fist 2 blocks) consists of 4 layers; a 3D downsampling *pooling* layer of filter size $2 \times 2 \times 2$ that shrinks the input volume to half of its size, a 3D convolutional layer of kernel $3 \times 3 \times 3$ with stride 1 and padding 1, a *Batch Normalization* [2] layer of momentum 0.95 for computing the running mean and variance and a *ReLU* non-linearity.

The settings for the network were fixed during the experiments. We used the Pytorch [9] framework for the training with batch size of 4 cubes per GPU and an initial learning rate 0.001 that decays by half every 20 epochs. The training was parallelized on 2 NVIDIA GPU Quadro RTX 5000. For the learning and weight optimization we use Adam optimizer [3] since it implies an adaptive learning rate for each layer after each step of optimization hence a faster convergence. To ensure that overfitting is avoided, we regularize our loss function using $L2$ regularization with penalty value 0.0001. We use a weighted *cross entropy loss* with the weights compensating for the low-sample classes more than those of higher sampled classes.

To train our networks, we choose the optimal model in a network optimization scheme. We use a validation set and we manually retrieve the model with the least loss and the highest accuracy. An example of the evaluation of the loss and accuracy is shown in Fig. 3, where results were obtained after the model converges without over-fitting.

3   Experiments

We perform several experiments. In the first one we study the direct classification of a vertebra in its 24 possible classes. In a second experiment we study a two-stage aggregated classification scheme. First, the anatomic group classification is performed (cervical, thoracic, lumbar), i.e. a 3-level classification problem. Then, three specialized networks are trained, one for each anatomic group.

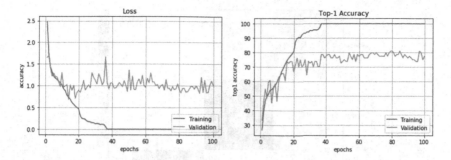

Fig. 3. Evaluation of the loss and accuracy of the training and validation sets for the 24 class classification optimization, w.r.t the number of epochs in training. First 8-fold.

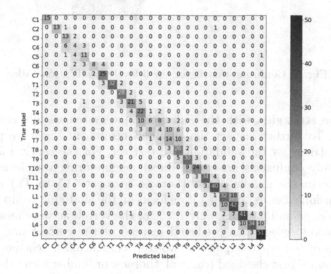

Fig. 4. Confusion matrix for the 24-level classification task.

3.1 24-level Classification

We first train our model to classify, individually, between all 24 vertebrae classes. We use all augmentation strategies presented in Sect. 2.2 and we use a random split of the data into train, validation and test. The evaluation of the loss and accuracy on the train and validation sets is presented in Fig. 3. The obtained classifications results have an accuracy of 80% on the test set. In order to assess the relevance of the different augmentations we performed the same experiment by only using one augmentation strategy. We present the results on Table 1, where *Rot* is only using the rotation augmentation, *Trans* is only using the translation transformation, *Rot + Trans* combines random transformations for each strategy on each sample, and *Full* includes all previous augmentations with scale and the noise augmentation. In all cases each sample was augmented with 10 random variations. The results on Table 1 show the benefit of the different

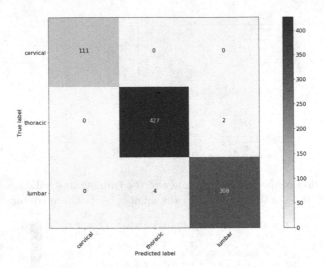

Fig. 5. Confusion matrix for the 3-level classification task.

augmentation strategies, with the *Full* technique systematically obtaining the best results. To further inspect the obtained results, in Fig. 4 we present the confusion matrices for the validation and test set. For the cervical and lumbar group, the only confusions arise with direct neighbouring vertebrae. However, the vertebrae in the middle of the thoracic segment (T5 to T9) present the highest confusions. Let us note that if we focus on the anatomic group predictions (cervical, thoracic and lumbar), only 1.5 % of the vertebrae were miss-classified, i.e. one anatomic group was predicted instead of the correct one. This high accuracy motivated us to explore a two-step classification scheme, where the anatomic group is first classified (cervical, thoracic or lumbar), and then a group-specific network is used.

3.2   3-level Classification

In the next experiment we train the same architecture network on the full augmented dataset but to only predict 3 labels, i.e. from a segmentation mask the network predicts an anatomic group: cervical, thoracic or lumbar.

We evaluate the full dataset using a 8-fold strategy and report the results on the aggregation of all test sets. The model achieves a 99.3% accuracy and the confusion matrix is presented in Fig. 5. These results show that the morphology of an individual vertebra contains relevant information to accurately distinguish the anatomic group of the vertebrae. Our next step is to study if a per-group model can better classify the individual vertebrae.

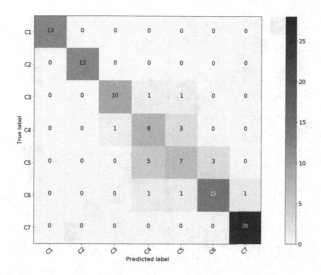

Fig. 6. Confusion matrix for the cervical vertebrae network.

3.3   Individual-per-group Classification

In this experiment, we train 3 per-group specific models, namely a model for the cervical vertebrae (7-classes), one for thoracic vertebrae (12-classes) and another model for the lumbar vertebrae (5-classes).

For the cervical group we obtain an accuracy of 84.68%. Figure 6 illustrates the confusion matrix of the cervical predictor. As expected, C1 and C2 have a very characteristic shape, making them for the network to be accurately identified. Then, we can observe that most confusions happen between C4, C5 and C6, while the shapes of C3 and C7 seem to be more accurately distinguishable with less confusions.

For the thoracic group we obtain an accuracy of 76.92% . Figure 7 illustrates the confusion matrix of the thoracic predictor. Most confusions arise in the section between T5 and T9, where distant vertebrae up to two neighbours are wrongly predicted (T7 for T5, T8 for T6 or T7 for T9). These results indicate that the shape of the middle section of the thoracic vertebrae is most similar, making them less individually identifiable.

For the lumbar group we obtain an accuracy of 86.08% . Figure 8 illustrates the confusion matrix of the lumbar predictor. It is worth noting that the failures are evenly distributed with the direct neighbouring vertebrae. While L1 and L2 have a slightly higher confusion rate, the shape of L4 and L5 seems to be more accurately distinguishable with less confusions.

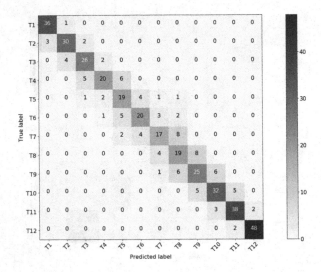

Fig. 7. Confusion matrix for the thoracic vertebrae network.

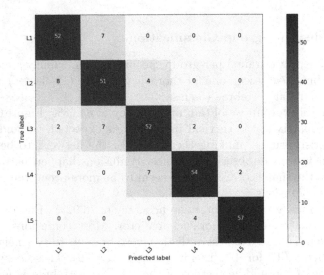

Fig. 8. Confusion matrix for the lumbar vertebrae network.

3.4   Comparison: 24-Level Vs 3-Level Plus Individual Per-Group

We studied two different approaches for the vertebrae classification: to directly predict one of the 24 vertebrae or to sequentially predict the anatomic group of a single vertebrae then use a specific per-group network.

The accuracy of the 24-level model is 71.07%, while the two steps aggregated accuracy is 82.56%.

A finer, per anatomic group analysis, shows that the 24-level model obtains an accuracy of 71.34% for the cervical vertebrae, 60.91% for the thoracic vertebrae and 75.14% for the lumbar vertebrae. For the two-stage method the results are 84.68% for the cervical vertebrae, 76.92% for the thoracic and 86.45% for the lumbar vertebrae.

This indicates that the two-stage approach is preferable in terms of classification accuracy, with a significant improvement in the classification of the cervical vertebrae. Interestingly, the trends in the confusion matrices in both approaches are consistent, with T5 to T9 presenting overall the greatest confusions.

4 Conclusion and Future Work

In this work we present a study on the classification of the vertebrae using its morphology. With the results of our current experiments we observe that the *individual* vertebrae classification is a non-trivial problem due to the morphological similarity of neighbouring vertebrae. However, our experiments confirm that classifying a vertebrae into its anatomic group is relatively straightforward for the networks. Moreover, our preliminary results point that: i) the shapes of the cervical and lumbar vertebrae seem to have enough information for a reliable individual identification, and ii) the thoracic vertebrae (T5 - T9) seem to have the highest similarity and are the ones where the network is most confused.

Our upcoming investigation will explore two leads: the use of neighbouring vertebrae to do the predictions and the exploration of other representations.

As most confusions arise with the network predicting the neighbouring vertebrae, we will consider an individual vertebrae with its surrounding neighbouring vertebrae as input. By considering the neighbouring vertebrae we hypothesize that the surrounding context will lead to a better classification. We plan to apply the same methodology: first we will refine the 3 anatomic classes into 5 classes(cervical only, cervical+thoracic, thoracic only, thoracic+lumbar, lumbar only). Then we will consider each individual class network as before: 7 for cervical vertebrae, 12 for thoracic vertebrae and 5 for lumbar vertebrae. Similarly we will compare the results of the two step method to the direct 24-level strategy.

A second lead of improvement will be to explore other representations for the input data. Instead of binary volumetric mask, we will consider *distance fields* [8] and *point clouds* [11,12]. By converting the data into different representations, we hypothesize that each representation may capture different features of the individual morphology.

While the VerSe 2019 Dataset contains some dysmorphic cases, the dataset does not have a medical label. Thus we did not assess the models performance on such anomalies. Future work will aim at quantifying the sensitivity of the trained models to pathological cases.

Acknowledgments. The work by Eslam Mohammed and Di Meng was funded by the SPINE PDCA project. The work by Sergi Pujades was funded by the ANR SEMBA project.

References

1. Camlica, Z., Tizhoosh, H.R., Khalvati, F.: Medical image classification via SVM using LBP features from saliency-based folded data. In: 2015 IEEE 14th International Conference on Machine Learning and Applications (ICMLA), pp. 128–132. IEEE (2015)
2. Ioffe, S., Szegedy, C.: Batch normalization: accelerating deep network training by reducing internal covariate shift. arXiv preprint arXiv:1502.03167 (2015)
3. Kingma, D.P., Ba, J.: Adam: a method for stochastic optimization. arXiv preprint arXiv:1412.6980 (2014)
4. Krizhevsky, A., Sutskever, I., Hinton, G.E.: ImageNet classification with deep convolutional neural networks. In: Advances in Neural Information Processing Systems, pp. 1097–1105 (2012)
5. Lessmann, N., van Ginneken, B., de Jong, P.A., Išgum, I.: Iterative fully convolutional neural networks for automatic vertebra segmentation and identification. Med. Image Anal. **53**, 142–155 (2019)
6. Liao, H., Mesfin, A., Luo, J.: Joint vertebrae identification and localization in spinal CT images by combining short-and long-range contextual information. IEEE Trans. Med. Imaging **37**(5), 1266–1275 (2018)
7. Lo, C.S., Wang, C.M.: Support vector machine for breast MR image classification. Comput. Math. Appl. **64**(5), 1153–1162 (2012)
8. Park, J.J., Florence, P., Straub, J., Newcombe, R., Lovegrove, S.: DeepSDF: learning continuous signed distance functions for shape representation. In: Proceedings of the IEEE Conference on Computer Vision and Pattern Recognition, pp. 165–174 (2019)
9. Paszke, A., et al.: PyTorch: an imperative style, high-performance deep learning library. In: Wallach, H., Larochelle, H., Beygelzimer, A., d'Alché-Buc, F., Fox, E., Garnett, R. (eds.) Advances in Neural Information Processing Systems, vol. 32, pp. 8024–8035. Curran Associates, Inc. (2019). http://papers.neurips.cc/paper/9015-pytorch-an-imperative-style-high-performance-deep-learning-library.pdf
10. Picuki: Vertebrae (2020). https://www.picuki.com/media/22278381395-84745065. Accessed 27 April 2020
11. Prokudin, S., Lassner, C., Romero, J.: Efficient learning on point clouds with basis point sets. In: Proceedings of the IEEE International Conference on Computer Vision Workshops (2019)
12. Qi, C.R., Su, H., Mo, K., Guibas, L.J.: PointNet: deep learning on point sets for 3D classification and segmentation. In: The IEEE Conference on Computer Vision and Pattern Recognition (CVPR), July 2017
13. Ronneberger, O., Fischer, P., Brox, T.: U-Net: convolutional networks for biomedical image segmentation. In: Navab, N., Hornegger, J., Wells, W.M., Frangi, A.F. (eds.) MICCAI 2015. LNCS, vol. 9351, pp. 234–241. Springer, Cham (2015). https://doi.org/10.1007/978-3-319-24574-4_28
14. Sekuboyina, A., et al.: Verse: a vertebrae labelling and segmentation benchmark. arXiv preprint arXiv:2001.09193 (2020)
15. Simonyan, K., Zisserman, A.: Very deep convolutional networks for large-scale image recognition. arXiv preprint arXiv:1409.1556 (2014)

Patient Specific Classification of Dental Root Canal and Crown Shape

Maxime Dumont[1], Juan Carlos Prieto[2], Serge Brosset[1], Lucia Cevidanes[1]([✉]),
Jonas Bianchi[1], Antonio Ruellas[1], Marcela Gurgel[1], Camila Massaro[1],
Aron Aliaga Del Castillo[1], Marcos Ioshida[1], Marilia Yatabe[1], Erika Benavides[1],
Hector Rios[1], Fabiana Soki[1], Gisele Neiva[1], Juan Fernando Aristizabal[3], Diego Rey[4],
Maria Antonia Alvarez[4], Kayvan Najarian[1], Jonathan Gryak[1], Martin Styner[2],
Jean-Christophe Fillion-Robin[5], Beatriz Paniagua[5], and Reza Soroushmehr[1]

[1] University of Michigan, 1011 North University Ave., Ann Arbor, MI 48109, USA
luciacev@umich.edu
[2] University of North Carolina, Chapel Hill, NC, USA
[3] University of Valle, Cali, Colombia
[4] University CES, Medellin, Colombia
[5] Kitware Incorporation, Clifton Park, USA

Abstract. This paper proposes machine learning approaches to support dentistry researchers in the context of integrating imaging modalities to analyze the morphology of tooth crowns and roots. One of the challenges to jointly analyze crowns and roots with precision is that two different image modalities are needed. Precision in dentistry is mainly driven by dental crown surfaces characteristics, but information on tooth root shape and position is of great value for successful root canal preparation, pulp regeneration, planning of orthodontic movement, restorative and implant dentistry. An innovative approach is to use image processing and machine learning to combine crown surfaces, obtained by intraoral scanners, with three dimensional volumetric images of the jaws and teeth root canals, obtained by cone beam computed tomography. In this paper, we propose a patient specific classification of dental root canal and crown shape analysis workflow that is widely applicable.

Keywords: Deep learning · Shape analysis · Dentistry

1 Introduction

In the context of dentistry imaging, machine learning techniques are becoming important to automatically isolate areas of interest in the dental crowns and roots [1]. Root resorption susceptibility has been associated to root morphology [2–4], and interest in variability in root morphology has increased recently [5–7]. Analysis of root canal and crown shape and position has numerous clinical applications, such as root canal treatment, regenerative endodontic therapies, restorative crown shape planning to avoid inadequate forces on roots and planning of orthodontic tooth movement. External apical

© Springer Nature Switzerland AG 2020
M. Reuter et al. (Eds.): ShapeMI 2020, LNCS 12474, pp. 145–153, 2020.
https://doi.org/10.1007/978-3-030-61056-2_12

root resorption (RR) is present in 7% to 15% of the population, and in 73% of individuals who had orthodontic treatment [8, 9].

Here, we propose automated root canal and crown segmentation methods based on image processing, machine learning approaches, shape analysis and geometric learning in medical imaging. The proposed methods are implemented in open source software solutions in two pipelines. The first pipeline is based on U-net for root canal and crown automatic segmentation from cone-beam computed tomography (CBCT) images and the second one is based on ResNet architecture for automatic segmentation of digital dental models (DDM) acquired with intraoral scanners. The proposed analysis is based on three main phases: (i) features extraction from raw volumetric and surface meshes, (ii) representative anatomic regions identification, and (iii) overall voxel-by-voxel and meshes vertices classification. The proposed automatic segmentation methods provide clinicians with labeling of the crown and root morphologies.

The paper sections are organized as follows. The next section reviews the materials and methods, describing the proposed approach. The experimental part and results are described in Sect. 3, followed by Sect. 4, dedicated to the discussion and conclusion.

2   Materials and Methods

2.1   Material

This retrospective study was approved by the Institutional Review Board. The sample of the present study was secondary data analysis and no CBCT scan was taken specifically for this research. The data consisted of 40 mandibular digital dental models (DDM) and CBCT scans for the same subjects. All subjects were imaged with the 2 imaging modalities. The mandibular CBCT scans were obtained using the Veraviewepocs 3D R100 (J Morita Corp.) with the following acquisition protocol: FOV 100 × 80 mm; 0.16 mm^3 voxel size; 90 kVp; 3 to 5 mA; and 9.3 s. DDM of the mandibular arch were acquired from intraoral scanning with the TRIOS 3D intraoral scanner (3 Shape; software version: TRIOS 1.3.4.5). The TRIOS intraoral scanner (IOS) utilizes "ultrafast optical sectioning" and confocal microscopy to generate 3D images from multiple 2-dimensional images with accuracy of 6.9 ± 0.9 μm. All scans were obtained according to the manufacturer's instructions, by 1 trained operator.

2.2   Methods

Two open-source software packages, ITK- snap, version 3.8 [10] and Slicer, version 4.11 [11] were used to perform user interactive manual segmentation of the volumetric images and common orientation of the mandibular dental arches for the learning model training. All IOS and CBCT scans were registered to each other using the validated protocol described by Ioshida et al. [12]. In this paper, we proposed two machine learning pipelines to boost the segmentation performance of root canal in volumetric images and dental crowns in surface scans. We first perform pre-processing in which we enhance the quality of images and increase the ratio between pixels belonging to root canals and background pixels . Then, we train a deep learning model to segment root canals. As there might be outliers in the results and the images might be over/under segmented we perform post-processing to address these issues.

Automatic Root Canal Segmentation. One of the main issues in many machine learning applications is class imbalance. To improve the accuracy in the first proposed pipeline and deal with the imbalance issue caused by low percentage of root canal pixels compared to the entire scan, we performed slice cropping to increase the ratio between the pixels which belong to root canals to the background pixels. All the 3D volumetric scans were cropped depending on their size in order to keep only the region of interest where the root canal pixels are. The cross-sectional images without root canals were automatically removed in order to feed the neural network almost exclusively with the images of interest. The algorithm selected the same anatomic cropping region for every 3D scan in the dataset, then split it into 2D cross-sections, and every cross-section was resized to 512×512 pixels. Contrast adjustment was performed as the original 3D scans had low contrast. After image pre-processing, 150 cross-sectional images were obtained for each patient.

UNet Model Training. This dataset was then trained in a UNet model [13, 14]. This network was first developed for biomedical image segmentation and then used in other applications, such as field boundary extraction from satellite images [15]. The network hierarchically extracts low-level features and recombines them into higher-level features in the encoder first. Then, it performs the element-wise classification from multiple features in the decoder. The encoder–decode architecture consists of down-sampling blocks to extract features and up-sampling blocks to infer the segmentation in the same resolution. The training has been done with 100 epochs, 400 steps per epochs, and a learning rate of 1e-5.

We also performed a 10-fold-cross validation in which each fold contains 4 scans. Therefore, 10 models were trained and for each model, 9 folds (36 scans) were used for training and 1 fold (4 scans) was used as the validation set. In order to identify the precision of the results, the quantitative measurements included Area Under the ROC Curves (AUC), F1 Score, accuracy, sensitivity and specificity were computed between the machine learning segmentation and the semi-manual segmentation. We did find a model that all of its measurements were the highest. Then the trained model with the highest AUC, and adequate F1 Score, accuracy, sensitivity and specificity, was selected and used it as the reference model for the segmentation.

Automatic Dental Crowns Segmentation. The data analytics of the 3D dental model surfaces requires extracting shape features of the dental crowns. The approach consists of taking 2D images or pictures of the 3D dental surface and extracting their associated shape features plus the corresponding label (background, gum, boundary between teeth and gum, teeth). The ground truth labeling of the dental surface is done with a region growing algorithm that uses the minimum curvature as stopping criteria plus manually correcting the miss-classified regions.

The 2D training samples are then generated by centering and scaling the mesh inside a sphere of radius one. The surface of the sphere is sampled regularly using an icosahedron sub-division approach, and each sample point is used to create a tangent plane to the sphere. The tangent plane serves as starting point of a ray-cast algorithm, *i.e.*, the dental surface is probed from the tangent plane. When an intersection is found, the surface normal and the distance to the intersection are used as image features. The corresponding ground truth label map is also extracted here.

Figure 1A shows the wireframe mesh of the sphere, the object inside the sphere, the tangent plane to the sphere and a perpendicular ray starting at the tangent plane. The plane resolution is set to 512 × 512. These purely geometric features are proposed because they led to higher accuracy in classification in our preliminary study [16], and do not depend on the position nor the orientation of the model, which may vary across the population. Each image is then used to train a modified Unet model with connections similar to a ResNet [17] training model. The modified architecture is shown in Fig. 1B and C. Specifically, the up-sampling block shown in Fig. 1B is modified to mirror the down-sampling blocks of ResNet in Fig. 1C, and this architecture is referred to as Ru-Net.

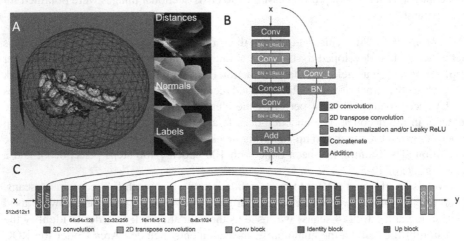

Fig. 1. PSCP meshes segmentation. A - Sphere and tangent plane created around the dental surface meshes and the shape features that are extracted; B - Up-sampling block; C - RUNET neural network architecture.

The prediction of a label on a new dental model is done following a majority vote scheme, *i.e.*, a single point on the dental surface may be captured by several tangent planes. The label with the greater number of votes is set as the final label for a specific point. A post processing step is applied to remove islands, *i.e.*, if a region has less than 1000 points, the label of the region is set to the closest labeled region. The final output of the algorithm is a labeled mesh with labels 0 as gingiva, 1 as boundary and 2 as dental crown. Finally, the calculated boundary helps provide individual labels for each crown. The final learning model, the DentalModelSeg, [18] was fed with 40 × 252 images/scan = 10,080 images. The DentalModelSeg tool, as part of the pipelines for patient specific classification and prediction (PSCP) tool, has been deployed in an open web-system for Data Storage, Computation and Integration, the DSCI [19], for execution of the automated tasks [20].

3 Results

3.1 Automatic Root Canal Segmentation

Quantitative measurements of AUC, F1 Score, accuracy, sensitivity and specificity are presented in Table 1. The F1 scores of the 10 folds presented a standard deviation of 0.077. As can be seen, the specificity values are higher than 0.99 with the standard deviation close to zero. However, the sensitivity values have higher standard deviation. The reason for getting low performance could be because of class imbalance and having low number of samples. Table 1 shows the average and standard deviation measurement for the 10 trained models and Fig. 2 shows an example of manual and automatic root canal segmentation.

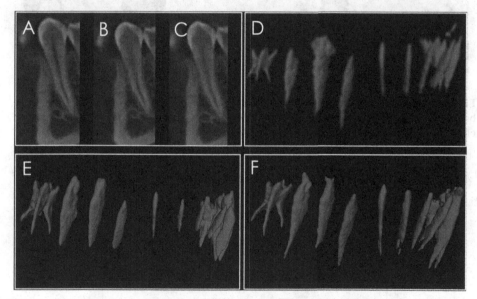

Fig. 2. A, CBCT scan gray level sagittal image at the premolar region; **B**, CBCT scan with manually labeled root canal; **C**, automatic segmentation combining image processing and machine learning approach; **D**, rendering of the root canals from molar to molar showing the initial automatic segmentation using image processing; **E**, manual segmentation, which often misses the apical portions of the root canal; and **F**, the combined image processing and machine learning segmentation that clearly identifies the root canal morphology for each tooth.

Table 1. AUC, F1 Score, accuracy, sensitivity and specificity of the proposed approach

	F1 Score	AUC	Sensitivity	Specificity	Accuracy
Average	0.7324	0.9174	0.8271	0.9997	0.9996
SD	0.0774	0.0548	0.1087	0.0001	0.0001

3.2 Automatic Dental Crowns Segmentation

The trained model detects a continuous boundary between the crown and the gingiva and segments the individual dental crowns for the 40 scans in the datasets, as shown in Fig. 3. The accuracy of the model is shown in a confusion matrix for random 7 scans (Table 2). The confusion matrix is a matrix where the diagonal component shows the ratio between the predicted label and the actual label. The closer it is to one, the better label was predicted there are. The 2nd label presented the worst performance (0.7) due to the dilation performed to detect boundaries.

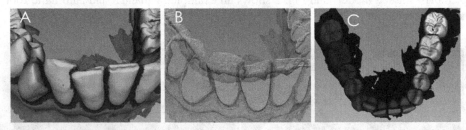

Fig. 3. A, the identification of shape parameters segments the dental crowns; **B**, errors in the boundaries are post-processed with region growing; and **C**, the machine learning segmentation clearly identifies the dental crown morphology for each tooth.

Table 2. Confusion matrix of the true and predicted labels for random 7 scans

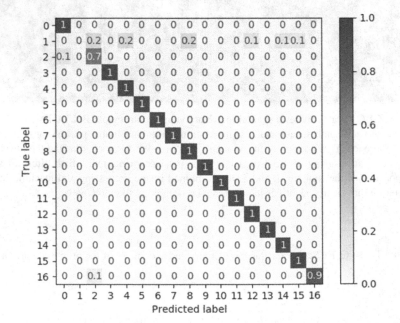

True \ Pred	0	1	2	3	4	5	6	7	8	9	10	11	12	13	14	15	16
0	1	0	0	0	0	0	0	0	0	0	0	0	0	0	0	0	0
1	0	0	0.2	0	0.2	0	0	0	0.2	0	0	0	0.1	0	0.1	0.1	0
2	0.1	0	0.7	0	0	0	0	0	0	0	0	0	0	0	0	0	0
3	0	0	0	1	0	0	0	0	0	0	0	0	0	0	0	0	0
4	0	0	0	0	1	0	0	0	0	0	0	0	0	0	0	0	0
5	0	0	0	0	0	1	0	0	0	0	0	0	0	0	0	0	0
6	0	0	0	0	0	0	1	0	0	0	0	0	0	0	0	0	0
7	0	0	0	0	0	0	0	1	0	0	0	0	0	0	0	0	0
8	0	0	0	0	0	0	0	0	1	0	0	0	0	0	0	0	0
9	0	0	0	0	0	0	0	0	0	1	0	0	0	0	0	0	0
10	0	0	0	0	0	0	0	0	0	0	1	0	0	0	0	0	0
11	0	0	0	0	0	0	0	0	0	0	0	1	0	0	0	0	0
12	0	0	0	0	0	0	0	0	0	0	0	0	1	0	0	0	0
13	0	0	0	0	0	0	0	0	0	0	0	0	0	1	0	0	0
14	0	0	0	0	0	0	0	0	0	0	0	0	0	0	1	0	0
15	0	0	0	0	0	0	0	0	0	0	0	0	0	0	0	1	0
16	0	0	0.1	0	0	0	0	0	0	0	0	0	0	0	0	0	0.9

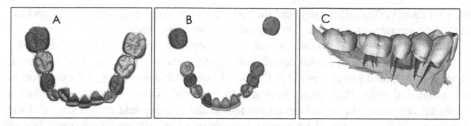

Fig. 4. The trained model was tested using in **A**, a mixed dentition DDM; and in **B**, a DDM with missing teeth; **C** shows the integration of root canal and crown morphology. Note that the trained model segmented the individual teeth in the mixed dentition in **A** properly, but segmentation of the DDM with missing teeth in **B** still requires further learning.

4 Discussion and Conclusions

Better understanding of the root canal morphology has the potential to increase the chance of successful root canal preparation, improve the treatment of pulpally involved teeth, and indicate the root position for planning both orthodontic movement and/or implant dentistry. In this paper we proposed automated image processing and machine learning based methods to segment both root canals and digital dental models automatically. The two different approaches presented in this paper facilitate the integration of the two imaging modalities, which provide complementary information to the dental clinician. The first approach uses U-net for automated root canal segmentation from CBCT images. The second approach uses ResNet architecture for automatic segmentation of crowns from digital dental models. As the same patients imaged with the CBCT were also imaged with the intraoral scanner, the multi-modal registration using a validated protocol facilitated the multi-modal fusion, consolidating the two approaches in service of one common patient.

In the automated root canal algorithm, we performed pre-processing to deal with the imbalance data issue and then trained deep learning models. While image processing performs well for in vitro root canal segmentations [21], as shown in Fig. 2D image processing alone often does not segment the root apex anatomy. The first model was based on U-Net architecture and the second one was based on RU-Net architecture. We performed 10-fold cross-validation and evaluated 10 models using different metrics including F1 score, AUC, sensitivity and specificity and achieved very high specificity. However, the sensitivity values need improvement. The low sensitivity performance could be due to variations in the CBCT scans field of view (e.g. some scans have only lower jaws and some of them have both jaws), class imbalance issues and small sample size. In our future work, to have more samples for training the models and improve the performance of the models, we will use the trained models to segment new scans first. Then, clinician users can interactively modify the segmented results, as it is easier to edit the segmented images rather than manually segment them from scratch.

Interestingly, the results shown in Fig. 2 reveal that the automatic segmentation can better identify the root canal apex compared to the manual segmentation. These segmented pixels on the deeper parts of the root canals increase the number of false positives, even though they belong to root canals. However, the manual segmentation of the root canals often fails to segment the root canal apex and anatomic details, and the automatic segmentation may in fact be more anatomically precise.

The training of the automatic dental crown segmentation in the DentalModelSeg tool was performed with digital dental models of permanent full dentition. All dental models were stored and computed in the DSCI open web-system [18] for execution of the automated tasks. Preliminary testing of the trained model included segmentation of dental crowns of the digital dental models in the primary and mixed stages of the dentition, as well as cases with unerupted, missing or ectopically positioned teeth. Future work will require individual labelling of the primary and/or permanent teeth as the individual labeling does not follow a specific order, and it is done based on the internal ordering of the points. Providing specific, unique labels for each crown is desired for generalizability of the proposed approaches (Fig. 4).

The automatic segmentation algorithms proposed in this study allow shape processing and analysis with precise learning and classification of whole tooth data. Both surface scanners images and grey level volumetric images can be segmented with accuracy with the methods presented in this paper. Analyzing and understanding 3D shapes for segmentation and classification remains a challenge due to various geometrical shapes of teeth, complex tooth arrangements, different dental model qualities, and varying degrees of crowding problems. Clinical applications of the proposed algorithms will benefit from future work comparing the performance of state-of-the-art neural networks and quantitative shape analysis of root and crown morphologies and position.

Acknowledgments. Supported by NIH DE R01DE024450, R21DE025306 and R01 EB021391.

References

1. Ko, C.C., et al.: Machine Learning in Orthodontics: Application Review. Craniofacial Growth Series, vol. 56, pp 117–135 (2020). http://hdl.handle.net/2027.42/153991
2. Xu, X., Liu, C., Zheng, Y.: 3D tooth segmentation and labeling using deep convolutional neural networks. IEEE Trans. Vis. Comput. Graph. **25**(7), 2336–2348 (2019). https://doi.org/10.1109/TVCG.2018.2839685
3. Elhaddaoui, R., et al.: Resorption of maxillary incisors after orthodontic treatment-clinical study of risk factors. Int. Orthod. **14**, 48–64 (2016). https://doi.org/10.1016/j.ortho.2015.12.015
4. Marques LS, Ramos-Jorge ML, Rey AC, Armond MC, Ruellas AC. Severe root resorption in orthodontic patients treated with the edgewise method: prevalence and predictive factors. Am J Orthod Dentofacial Orthop 2010; 137: 384 ± 8. https://doi.org/10.1016/j.ajodo.2008.04.024
5. Marques, L.S., Chaves, K.C., Rey, A.C., Pereira, L.J., Ruellas, A.C.: Severe root resorption and orthodontic treatment: clinical implications after 25 years of follow-up. Am. J. Orthod. Dentofac. Orthop. **139**, S166–S169 (2011). https://doi.org/10.1016/j.ajodo.2009.05.032
6. Kamble, R.H., Lohkare, S., Hararey, P.V., Mundada, R.D.: Stress distribution pattern in a root of maxillary central incisor having various root morphologies: a finite element study. Angle Orthod. **82**, 799–805 (2012). https://doi.org/10.2319/083111-560.1
7. Oyama, K., Motoyoshi, M., Hirabayashi, M., Hosoi, K., Shimizu, N.: Effects of root morphology on stress distribution at the root apex. Eur. J. Orthod. **29**, 113–117 (2007). https://doi.org/10.1093/ejo/cjl043

8. Lupi, J.E., Handelman, C.S., Sadowsky, C.: Prevalence and severity of apical root resorption and alveolar bone loss in orthodontically treated adults. Am. J. Orthod. Dentofac. Orthop. **109**(1), 28–37 (1996). https://doi.org/10.1016/s0889-5406(96)70160-9

9. Ahlbrecht, C.A., et al.: Three-dimensional characterization of root morphology for maxillary incisors. PLoS ONE **12**(6), e0178728 (2017). https://doi.org/10.1371/journal.pone.0178728

10. ITK- snap. www.itksnap.org(2020). Accessed 30 June 2020

11. Slicer, version 4.11. www.slicer.org. Accessed 30 June 2020

12. Ioshida, M., et al.: Accuracy and reliability of mandibular digital model registration with use of the mucogingival junction as the reference. Oral Surg. Oral Med. Oral Pathol. Oral Radiol. **127**(4), 351–360 (2019). https://doi.org/10.1016/j.oooo.2018.10.003

13. Ronneberger, O., Fischer, P., Brox, T.: U-net: convolutional networks for biomedical image segmentation. In: Navab, N., Hornegger, J., Wells, W., Frangi, A. (eds.) International Conference on Medical Image Computing and Computer-Assisted Intervention. Lecture Notes in Computer Science, vol. 9351, pp. 234–241. Springer, Cham (2015). https://doi.org/10.1007/978-3-319-24574-4_28

14. https://github.com/zhixuhao/unet. Accessed 30 June 2020

15. Waldner, F., Diakogiannis, F.I.: Deep learning on edge: ex-tracting field boundaries from satellite images with a convolutional neural network. Remote Sens. Environ. **245**, 111741 (2020)

16. Ribera, N.T.: Shape variation analyzer: a classifier for temporomandibular joint damaged by osteoarthritis. Proc SPIE Int. Soc. Opt. Eng. **10950**, 1095021 (2019). https://doi.org/10.1117/12.2506018

17. He, K., Zhang, X., Ren, S., Sun, J.: Deep residual learning for image recognition. In: Proceedings of the IEEE Computer Society Conference on Computer Vision and Pattern Recognition, vol. 7780459, pp. 770–778 (2016)

18. DentalModelSeg source code and documentation. https://github.com/DCBIA-OrthoLab/fly-by-cnn. Accessed 30 June 2020

19. Data Storage Computation and Integration, DSCI. www.dsci.dent.umich.edu. Accessed 30 June 2020

20. Michoud, L., et al.: A web-based system for statistical shape analysis in temporomandibular joint osteoarthritis. Proc. SPIE Int. Soc. Opt. Eng. **10953**, 109530T (2019). https://doi.org/10.1117/12.250603

21. Michetti, J., Basarab. A., Diemer, F., Kouame, D.: Comparison of an adaptive local thresholding method on CBCT and μCT endodontic images. Phys. Med. Biol. **63**(1), 015020 (2017). https://doi.org/10.1088/1361-6560/aa90ff

Author Index

Printed in the United States
By Bookmasters

Printed in the United States
By Bookmasters